Blockchain for IoT

Blockchain for IoT provides the basic concepts of Blockchain technology and its applications to varied domains catering to socio-technical fields. It also introduces intelligent Blockchain platforms by way of infusing elements of computational intelligence into Blockchain technology. With the help of an interdisciplinary approach, it includes insights into real-life IoT applications to enable the readers to assimilate the concepts with ease. This book provides a balanced approach between theoretical understanding and practical applications.

Features:

- A self-contained approach to integrating the principles of Blockchain with elements of computational intelligence.

- A rich and novel foundation of Blockchain technology with reference to the internet of things conjoined with the tenets of artificial intelligence in yielding intelligent Blockchain platforms.

- Elucidates essential background, concepts, definitions, and theories thereby putting forward a complete treatment on the subject.

- Information presented in an accessible way for research students of computer science and information technology, as well as software professionals who can inherit the much-needed developmental ideas to boost up their computing knowledge on distributed platforms.

This book is aimed primarily at undergraduates, postgraduates, and researchers studying Blockchain.

Chapman & Hall/CRC Computational Intelligence and Its Applications

Series Editor: *Siddhartha Bhattacharyya*

Intelligent Copyright Protection for Images
Subhrajit Sinha Roy, Abhishek Basu, Avik Chattopadhyay

Emerging Trends in Disruptive Technology Management for Sustainable Development
Rik Das, Mahua Banerjee, Sourav De

Computational Intelligence for Human Action Recognition
Sourav De, Paramartha Dutta

Disruptive Trends in Computer Aided Diagnosis
Rik Das, Sudarshan Nandy, Siddhartha Bhattacharyya

Intelligent Modelling, Prediction, and Diagnosis from Epidemiological Data: COVID-19 and Beyond
Siddhartha Bhattacharyya

Blockchain for IoT
Debarka Mukhopadhyay, Siddhartha Bhattacharyya, Balachandran Krishnan, Sudipta Roy

For more information about this series please visit:
https://www.crcpress.com/Chapman--HallCRC-Computational-Intelligence-and-Its-Applications/book-series/CIAFOCUS

Blockchain for IoT

Edited by
Debarka Mukhopadhyay
Siddhartha Bhattacharyya
Balachandran Krishnan
Sudipta Roy

CRC Press
Taylor & Francis Group
Boca Raton London New York

CRC Press is an imprint of the
Taylor & Francis Group, an **Informa** business
A CHAPMAN & HALL BOOK

First edition published 2023
by CRC Press
6000 Broken Sound Parkway NW, Suite 300, Boca Raton, FL 33487-2742

and by CRC Press
4 Park Square, Milton Park, Abingdon, Oxon, OX14 4RN

CRC Press is an imprint of Taylor & Francis Group, LLC

Library of Congress Cataloging-in-Publication Data
Names: Mukhopadhyay, Debarka, editor.
Title: Blockchain for IoT / edited by Debarka Mukhopadhyay, Siddhartha Bhattacharyya, Balachandran Krishnan, Sudipta Roy.
Description: First edition. | Boca Raton : Chapman & Hall/CRC Press, 2023. | Series: Chapman & Hall/CRC computational intelligence and its applications | Includes bibliographical references and index. Identifiers: LCCN 2022018412 (print) | LCCN 2022018413 (ebook) | ISBN 9781032036212 (hbk) | ISBN 9781032036243 (pbk) | ISBN 9781003188247 (ebk)
Subjects: LCSH: Blockchains (Databases) | Internet of things.
Classification: LCC QA76.9.B56 B557 2023 (print) | LCC QA76.9.B56 (ebook) DDC 005.74--dc23/eng/20220805
LC record available at https://lccn.loc.gov/2022018412
LC ebook record available at https://lccn.loc.gov/2022018413

ISBN: 978-1-032-03621-2 (hbk)
ISBN: 978-1-032-03624-3 (pbk)
ISBN: 978-1-003-18824-7 (ebk)

DOI: 10.1201/9781003188247

Typeset in Minion
by SPi Technologies India Pvt Ltd (Straive)

Contents

Preface

In today's world, Internet of Things (IoT) industries from various geographical positions communicate via several organizational contracts. This is nothing but a formal agreement depending on mutual trust between participating industries. The Blockchain platform establishes this trust with the help of technology. Thus there is a fundamental difference between existing communication and communication through Blockchain technology. One of the key goals of this work is to showcase Blockchain as a platform for transparent and real-time information interchange between the participants. Moreover, computational intelligence-enabled features are incorporated with the existing properties of Blockchain and IoT systems for better utilization of the platform.

This volume aims to introduce the basic concepts of Blockchain technology and its applications to varied domains to the intended readers. The book also intends to introduce intelligent Blockchain platforms by way of infusing elements of computational intelligence. Insights into real-life IoT applications would help the readers assimilate the concepts with ease. The volume comprises nine contributory chapters in addition to the introductory chapter to report the latest developments in this field.

Blockchain is a decentralized or distributed platform for sharing information and computational capabilities. Records of transactions are duplicated among the participating workstations in the network, which coordinate and collaborate with each other. As such, any attempt to change the stored data requires access to all the workstations in the network, thereby ensuring higher secrecy and security of the underlying data. IoT refers to the connectivity of numerous physical objects or devices (either sensors or actuators) over the Internet to ascertain a densely connected network infrastructure capable of making decisions on its own. Thus, a huge amount of sensitive data needs to be processed and stored for further use in an IoT network. Given its robust security, Blockchain finds widespread applications in several fields of applications ranging from science and engineering to medical and business applications, and the IoT is no exception in this regard. Chapter 1 introduces the fundamentals of Blockchain and the IoT with special reference to the synergy between the two in yielding a robust and secure distributed data-processing environment.

Blockchain is one of the latest and rising technologies of the decade. When bitcoin and the idea of cryptocurrency were first introduced to the market in 2008, some saw the revolutionizing potential of Blockchain technology, the platform behind cryptocurrency. Blockchain technology helps businesses provide security to data and information, reduces costs by eliminating unnecessary middlemen, and delivers transparency and traceability. This ensures end-to-end synchronization, i.e., having a perspective from the customer's end. Blockchain is a public ledger that maintains records of all the transactions held on a Blockchain network while working in a distributed manner. It is a peer-to-peer (P2P) network that does not require any central authority. The global market for Blockchain technology is projected to reach US\$30.7 billion by 2027, trailing a post-COVID-19 compound annual growth rate (CAGR) of 43% over the analysis period 2020 through 2027. IoT is another rapidly growing technology that has captured the attention of many enterprises by its ability to deliver a smart network of connected devices that can enhance the user's experiences in interacting with a device. IoT is a growing industry and has been paving the way for the Industry 4.0 revolution. The limitations demonstrated by IoT in the real world have not deterred researchers and enterprises from finding solutions. One of the most anticipated upcoming solutions is the integration of Blockchain with IoT. Chapter 2 identifies various applications of IoT on the Blockchain platform and the associated challenges/benefits compared to the traditional system. The study aims to determine the challenges faced by the Blockchain technology with IoT as a technology and its possible solutions.

At present, extensive implementation of IoT technologies is resulting in smarter devices with the rapid growth and widespread application. On the other hand, it has encountered some serious security concerns. Device and message validation in addition to verification using an authentic mechanism is crucial among the distributed nodes in the network to prevent access by unauthorized devices, among other security issues such as data privacy, attack resistance, and so on. Many IoT settings, on the other hand, are vulnerable to a node or device failure. To prevent single data node failure and data tampering, Blockchain technology could be used to archive the distribution of attributes. A generalized Blockchain-based scheme has been explored in Chapter 3 that focuses on authentication and key management mechanisms using public-key cryptographic algorithms. An exhaustive analysis of integrating Blockchain with IoT authentication is reviewed to improve Blockchain-based IoT applications, highlighting its effects in a conventional distributed cloud.

The IoT and Blockchain knowledge benefits the implementing organizations to enhance and optimize their optimization. These days, the IoT has drawn immense interest from scholars, scientists, and businesspeople because of its ability to offer inventive administrations across different applications. IoT flawlessly interconnects heterogeneous gadgets and objects to make an actual organization in which detecting, handling, and correspondence procedures are naturally organized and overseen without human intervention. Distinctive organization advances include cyber-physical systems (CPS), machine-to-machine (M2M), and wireless sensor networks (WSNs). As present-day technologies upscale, there is a need for us to push the legacy systems toward the convergence of IoT and Blockchain (IoTBC). Chapter 4 discusses some of the smart frameworks that have resulted from such convergence. The world today has completely accustomed itself to the existing legacy systems, making them difficult to replace. Thus, security concerns identifying with CPS, M2M, and WSN emerge in related fields, which require ensuring the safety of organization systems as a whole against assaults. The security part of the system could be easily handled with respect to the Blockchain technology. The IoTBC merger brings numerous benefits, and it is explained in reference to the simple voice calling system. Future systems such as smart healthcare, governance, and infrastructure also are discussed in detail. iGovernance is tied in with utilizing innovation to work with and support better preparation and navigation. It is tied in with working on just cycles and changing the manner in which public administration efforts are conveyed. The significant parts of medical services frameworks are distinguishing proof, area, detecting, and network. iHealthcare is carried out through a wide scope of frameworks: crisis administrations, brilliant processing, sensors, lab on chips, remote checking, wearable gadgets, availability gadgets, and large information. The IoT-based frameworks are furnished with body sensor networks inside telemedicine frameworks.

In recent times, the smart home automation system is one of the fastest emerging IoT technologies because of the availability of remote access facilities by using which one can control devices and improve the standard of living. In Chapter 5, an IoT-based smart home automation system using a wireless fidelity shield and Arduino microcontroller to control household electrical appliances remotely is proposed. This system can help old-age and physically challenged people operate the conventional electrical switches fixed somewhere in the house. Moreover, the system can provide a comfortable and independent lifestyle with minimum cost.

In previous studies on bitcoin, it is noted as a vision of a new decentralized economic system. Financial institutions are not included in solely peer-to-peer electronic cash transactions, with online payments sent directly from one individual to another. The practical applicability of this concept is discussed in Chapter 6, which presents a practical real-time study of a decentralized autonomous organization (DAO) automatically driven with full-proof transparency, integrity, and decentralization. Decentralization means that it is executed in the peer-to-peer format including nodes in the network for users, and it is fully transparent because the network is open in the public domain. All these characteristics are taken together in this chapter to form the milestones of trust-free transactions systems.

The Blockchain has already modernized how digital assets are shared among users by providing an immutable trace of their transactions. The power of artificial intelligence (AI) and the Blockchain can help in reforming legal systems by ensuring reliable execution of legitimate contracts, reducing costs, and speeding up judicial conclusions. For example, a legal career involves the reading of several thousand legal texts that are important in deciding the outcome of a trial. Today, many automatic chatbots are designed with advanced natural language processing algorithms that can read these texts in minutes. Automation of guaranteed compliance of legal contracts is one of the vital tasks in the legal industry, which can be reliably implemented using smart contracts. Blockchain and AI can be effectively utilized in several areas to support the varying nature of legal work to facilitate novel business areas with multiple service offerings. Chapter 7 discusses the impact of Blockchain in various legal industry tasks such as intellectual property rights, chain of custody, individual property rights, and notary public verification. It also presents a practical scenario of judgment execution on Blockchain and explains various natural language processing techniques and their challenges in legal text analytics.

An IoT network uses a limited number of resource-constrained devices for its application, which is not compatible with performing satisfactorily in the presence of dynamic changes in the network behavior. Due to the resource-constrained nature of IoT networks, the support for powerful high-end infrastructure is not feasible. Providing adequate security

with limited resources becomes a challenging task for maintaining the authenticity and integrity of the data. To overcome the issue of data authenticity and integrity in IoT networks, the authors propose a software-defined network (SDN)-based agent for Blockchain-based IoT networks in Chapter 8. The proposed solution is an SDN-based IoT architecture that uses an agent powered by machine learning to sustain the rapid change and erroneous behavior of the IoT network. Compared to the current state-of-the-art approach, the proposed system will provide enhanced performance due to its agent-based adoptability feature that can work well with the dynamic changing behavior of the IoT network.

Distributed generation, energy storage, and other IoT devices (such as smart meters) are working together to generate new opportunities for centralized control operations and energy markets. Blockchain provides businesses and IoT devices with a potential route to better navigate these changes by optimizing the use of ever-increasing volumes of energy data and providing organizations and individuals with new ways of transacting and building a trustworthy energy transaction management network without the need for a central authority. Chapter 9 provides a detailed illustration of the state-of-the-art IoT-enabled peer-to-peer (P2P) trading of rooftop solar (RTPV) power on a Blockchain platform in India. Using IoT-enabled smart meters, P2P trading systems provide a virtual platform where prosumers and consumers may trade electricity at a mutually agreed-upon price without the need for an intermediary. India Smart Grid Forum (ISGF), in collaboration with Power Ledger, has demonstrated two successful first-of-their-kind pilots in Southeast Asia in India. The first pilot project in Lucknow, Uttar Pradesh, went live in December 2020 with 12 participants, and Tata Power Delhi Distribution Ltd. (TPDDL) went live in January 2021 with over 65 prosumers and 75 consumers. Based on the various trading rules that have been used since the pilots started, it has been demonstrated that a network access charge for the DISCOM may be included in this type of trading scenario, incentivizing the DISCOM to allow P2P energy trading among their customers while also promoting widespread adoption of smart meters across the country.

Agriculture is fundamental for the survival of humankind. To cater to the current needs and decrease food wastage, smart agriculture using innovative technologies like the IoT, AI, and Big Data is the current trend. Blockchain technology is recently being used to solve the challenges of transparency, neutrality, security, and reliability in agriculture. In Chapter 10, the authors present a clear picture of smart agriculture using Blockchain, IoT, and AI together. A framework for smart agriculture is proposed wherein IoT devices will collect data related to agricultural production, transport, and storage. AI is used to gather specific knowledge from collected data and for future prediction. Blockchain is used at every step of storage of data after sensing or transaction, and all the authorized parties can verify data from the Blockchain. In comparison with existing methods, the proposed framework is found to exhibit more features in many parameters.

This book covers a strong foundation of Blockchain as a platform together with the application of AI for the IoT industry incorporating real-life case studies. It covers various topics such as basic to most advanced Blockchain platforms, implementations of sophisticated artificial intelligence concepts, etc. This book provides a balanced approach between theoretical understanding and practical applications. The primary audience of this book includes the research students of computer science and information technology, as well as software professionals who can use the much-needed developmental ideas to boost up their computing knowledge on distributed platforms.

Debarka Mukhopadhyay
Siddhartha Bhattacharyya
Balachandran Krishnan
Sudipta Roy

Editors

Dr. Debarka Mukhopadhyay has more than 17 years of research and teaching experience and is currently Associate Professor in the Department of Computer Science and Engineering at Christ University, Bangalore. He is an enthusiastic and result-oriented individual with a Ph.D. in Computer Science & Engineering from Maulana Abul Kalam Azad University of Technology, West Bengal, and M. Tech in Computer Science and Engineering from Kalyani Govt Engineering College under the West Bengal University of Technology. Dr. Mukhopadhyay holds 20 patents including 6 international and 14 national (Indian), and published 14 articles in international journals, 22 in conference proceedings, as well as 8 book chapters. His current research interests include Blockchain technology, IoT, AI, and nano computing, among others. He is a member of the Institute of Electrical and Electronics Engineers (IEEE), the Indian Society for Technical Education (ISTE), and the Institution of Electronics and Telecommunication Engineers (IETE).

Dr. Siddhartha Bhattacharyya [FRSA, FIET (UK), FIEI, FIETE, LFOSI, SMIEEE, SMACM, SMIETI, SMAAIA, LMCSI, LMISTE] earned a Bachelor's degree in Physics in 1995, a Bachelor's degree in Optics and Optoelectronics in 1998, and a Master's degree in Optics and Optoelectronics in 2000, all from the University of Calcutta, India. He completed a

Ph.D. in Computer Science and Engineering from Jadavpur University, India in 2008. He is the recipient of the University Gold Medal from the University of Calcutta for his Master's thesis. He is the recipient of several coveted awards including the Distinguished HoD Award and Distinguished Professor Award conferred by the Computer Society of India, Mumbai Chapter, India in 2017, the Honorary Doctorate Award (D. Litt.) from The University of South America, and the South East Asian Regional Computing Confederation (SEARCC) International Digital Award ICT Educator of the Year in 2017. He has been appointed as the ACM Distinguished Speaker for the 2018–2020 tenure. He has been inducted into the People of ACM Hall of Fame by ACM, USA, in 2020. He has been appointed as the IEEE Computer Society Distinguished Visitor for the 2021–2023 tenure. He has been elected as the full foreign member of the Russian Academy of Natural Sciences. He has been elected a full fellow of The Royal Society for Arts, Manufacturers and Commerce (RSA), London, UK. Dr. Bhattacharyya is currently serving as the Principal of Rajnagar Mahavidyalaya, Rajnagar, Birbhum. He is a coauthor of 6 books and coeditor of 80 books and has more than 350 published research publications in international journals and conference proceedings to his credit. He is the owner of 2 patents under the Patent Cooperation Treaty (PCT) and 19 patents. He has been a member of the organizing and technical program committees of several national and international conferences. Dr. Bhattacharyya is an associate editor of several reputed journals including *Applied Soft Computing, IEEE Access, Evolutionary Intelligence,* and *IET Quantum Communications.* He is the editor of the *International Journal of Pattern Recognition Research* and the Founding Editor in Chief of the *International Journal of Hybrid Intelligence, Inderscience.* He has guest-edited several issues with several international journals. His research interests include hybrid intelligence, pattern recognition, multimedia data processing, social networks, and quantum computing.

Dr. Balachandran Krishnan is a professor in the Department of Computer Science and Engineering, Christ University, Bangalore. He earned his Ph.D. in Information and Communication Engineering from Anna University, Chennai, Tamil Nadu. He did his postgraduate work in molecular physics, computer application, and information technology. His research includes developing an adaptive weight-based ensemble model for lung cancer pre-diagnosis. His research interests include artificial intelligence, data science, and computer networks. In his 38 years of professional experience, he has worked as a scientific officer in the atomic energy department of a research and development center catering to assay and analysis of geophysical data exploration of radioactive minerals (20 years) and as a field officer with the Tata Tea Plantation (1 year), in addition to his current tenure at Christ University. He is guiding research scholars in the area of data science, image processing, and software engineering.

Sudipta Roy, Ph.D. is a Professor in the Department of Computer Science and Engineering, Triguna Sen School of Technology, Assam University, Silchar, Assam. He was associated with ACE Consultants, Kolkata, as a software professional, and Bengal Institute of Technology, Kolkata, India, as a faculty member. His research interests include image processing, soft computing techniques, and network security. He is presently guiding research scholars as well as postgraduate and graduate students. He has published more than 80 papers in international journals and more than 70 in conference proceedings. He is a member/fellow of Institution of Engineers (IE), Institution of Electronics and Telecommunication Engineers (IETE), Computer Society of India (CSI), Institute of Electrical and Electronics Engineers (IEEE), International Neural Network Society (INNS), and Association for Computing Machinery (ACM).

Contributors

Tapodhir Acharjee
Department of Computer Science
& Engineering
Triguna Sen School of Technology
Assam University, Silchar
Assam, India

Ayan Bhattacharjee
Tripura University (A Central
University)
Tripura, India

Siddhartha Bhattacharyya
Rajnagar Mahavidyalaya
Birbhum, India

Rohit Kumar Das
VIT-AP University
Andhra Pradesh, India

Sajal Kanta Das
Tripura University (A Central
University)
Tripura, India

Raktim Deb
Department of Computer Science
& Engineering
Triguna Sen School of Technology
Assam University, Silchar
Assam, India

Mampi Devi
Tripura University(A Central
University)
Tripura, India

Rohit Kumar Kasera
Department of Computer Science
& Engineering
Triguna Sen School of
Technology
Assam University, Silchar
Assam, India

Balachandran Krishnan
Christ University
Bangalore, India

G. Megala
School of Computer Science and
Engineering
Vellore Institute of Technology,
Vellore
Tamil Nadu, India

Debarka Mukhopadhyay
Christ University
Bangalore, India

Julian Benadit Pernabas
Department of Computer Science
   and Engineering, School of
   Engineering and Technology
CHRIST (Deemed To be
   University)
Kengeri Campus, Kanmanike,
   Bangalore, Karnataka, India

Vandana Reddy
CHRIST Deemed to be University
Bangalore, India

Alak Roy
Tripura University (A Central
   University)
Tripura, India

Shuvam Sarkar Roy
The World Bank
New Delhi, India

Sudipta Roy
Assam University, Silchar
Assam, India

Jitendra Saxena
The Institution of Engineers
Kolkata, India

Prabu Sevugan
Department of Banking
   Technology
Pondicherry University
Pondicherry, India

P. Swarnalatha
School of Computer Science and
   Engineering
Vellore Institute of Technology,
   Vellore
Tamil Nadu, India

Krishna Kumar Vaithinathan
Department of Computer
   Engineering
Karaikal Polytechnic College,
   Varichikudy
Karaikal, Puducherry, India

Archana Yengkhom
Christ University
Bangalore, India

CHAPTER 1

Introduction to Blockchain for Internet of Things

Debarka Mukhopadhyay
Christ University, Bangalore, India

Siddhartha Bhattacharyya
Rajnagar Mahavidyalaya, Birbhum, India

Sudipta Roy
Assam University, Silchar, Assam, India

Balachandran Krishnan
Christ University, Bangalore, India

CONTENTS

DOI: 10.1201/9781003188247-1

1.1 INTRODUCTION

Since about the 20th century, information—or data—is treated as one of the most valuable parts of any organization. Data can be very important, and keeping secure access to data is essential. Hence a requirement for a highlysecured architecture platform for data storage [1]. A centralized system is not ideal because it represents a single point of failure. If we turn to a decentralized system, in case of any failure of a node, a lot of time may be required to reconstruct the communication [2]. By comparison, Blockchain as a distributed framework guarantees fidelity, privacy, and security for data and moreover creates trust without involvement of any trusted third party. The distributed database is shared among all the participants in the network. Blockchain is best known as a platform for cryptocurrencies like Bitcoin, Ether, etc. A conventional database differs from a Blockchain database [3]. In Blockchain, data are stored in containers known asa block. Blocks have varied capacity, and when each one is filled, it is closed and then linked with a previous block, thus creating a chain. Once a new block is created and added to the existing chain, it becomes immutable and data within it cannot be altered. Hence no new data can be incorporated inside the existing blocks in the chain, but can be a part of the next block. New blocks can be added to an existing chain as long as more data storage is required [4].

So Blockchain allows data to be distributed but not edited. Hence, any transactions that are recorded cannot be altered, modified, or destroyed. Now, let us assume that a hacker wants to hack information from the chain. If successful, they own a node in the chain [5]. If they want to modify their own copy, they won't be in alignment with other nodes, as the changes will be reflected everywhere and every participant can identify the single modified copy; the hacked node is thus deemed illegitimate and is cast away. For a successful hack of a Blockchain, the intruder must control a minimum of 51% of the nodes. But this requires a huge amount of time, effort, and computational power required to redo the timestamps and hash values.

While talking about the features, this framework contributes to a lot of areas. The Blockchain framework can be implemented to transfer money from one geographical location to another, especially cross-border transfer. Such transfers can take days as a traditional bank transaction, whereas a blockchain platform can facilitate it in minutes. This distributed platform does not require the participants to deposit money to any centralized authority for any transaction. Though it mainly deals with cryptocurrencies, the concept may be incorporated into traditional investments as well [6].

Money lending can be made possible using smart contracts. A smart contract is an algorithm that resides in Blockchain and allows some defined activities like processing and various types of payment to trigger automatically. It makes processing faster and less expensive, as no third-party involvement is necessary [7].

The insurance sector can use Blockchain to bring higher transparency between the insurance provider and the customer. A smart contract built into a Blockchain payment architecture can speedup payments in case of claims [8].

Real estate sector requires a lot of paperwork to initiate the process of searching for existing ownership to transfer deeds and titles. A Blockchain platform can keep all such records for verification to transfer ownership. This process can reduce time, paperwork, and costs.

Blockchain technology can improve current voting systems. It can check eligibility of an individual as voter and flag any duplication. Moreover, the eligible electors will be able to cast votes from anywhere, which will significantly reduce the costs of conducting an election [9].

Another application is in the field of supply chain management [10]. Blockchain technology can bring transparency in communication between the participants in a supply chain. It also ensures data security and integrity, as data in a Blockchain is immutable. Thus, the partners in a supply chain can have more trust in one another, as well as in the data being updated appropriately.

Blockchain technology has other areas of applications as well. With growing needs of digital data, Blockchain can ensure data security, transparency, and integrity.

1.2 HISTORY

The concept of Bitcoin came into existence in 2008 and was originally proposed by a person—or possibly a group of people—known as Satoshi Nakamoto. But even before Nakamoto published the while paper on Bitcoin, the concept of blockchain technology has already appeared in earlier research. The preliminary concept came into existence with Stuart Haber and W. Scott Stornetta in 1991. The work proposed a cryptographically secured chain with time-stamped documents [11]. They proposed ways to determine hash values of documents and saving them with a time stamp. Nick Szabo, a computer scientist, legal scholar, and cryptographer, conceptualized digital currency in 1998, then known as Bit Gold. This proposal incorporated solving cryptography puzzles, Byzantine fault-tolerant registry, and validating and time-stamping a newly "mined" coin. He also tried to solve the problem of double spending. Bit Gold was never implemented, but the concept was utilized in Bitcoin architecture. In 2000, Stefan Konst introduced a theory of connecting blocks in a chain. He implemented this theory with graphs where nodes are linked using a cryptography method. Nakamoto eventually compiled past theories and in 2009 implemented Bitcoin, "A peer to peer version of electronic cash" [11].

1.3 WHAT IS A BLOCKCHAIN?

Blockchain is a decentralized or distributed platform for information sharing and computation. Records of transactions are duplicated among the participating workstations in the network that do not trust each other but coordinate and collaborate. Unlike centralized system, here possible attackers need to access all the workstations in the network to change similar data. This action of attackers is reported immediately among the participants within the network due to inherent characteristics of Blockchain technology [12]. Figure 1.1 identifies various characteristics of Blockchain as a platform. It offers open participation of a workstation without revealing its identity. Like any other distributed or decentralized network, this platform also exhibits no centralized control over the transactions. Whenever a transaction is initiated between any two workstations, the other workstations do not remain idle but try to explore the validity of the transaction initiated by the workstations. Thus, the remaining workstations perform as validators for the

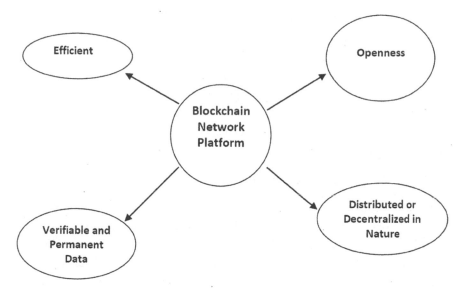

FIGURE 1.1   Different characteristics of Blockchain network platform.

transaction. After successful validation and mining, the information is stored permanently within storage [13].

1.4 BLOCKCHAIN AS PUBLIC LEDGERS

In a distributed peer-to-peer network, each workstation houses a dedicated database that contains an identical copy of a ledger and gets updated simultaneously and independently. Like all public ledgers, a Blockchain network keeps records as a record book of all successful transactions between participants. Figure 1.2 depicts fundamental differences between

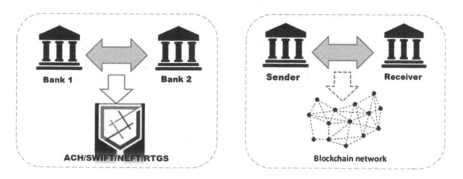

FIGURE 1.2   Regular fund transfer versus transfer through Blockchain [14].

conventional and Blockchain-based transactions. In banking system, for example, a sender can transfer funds to a receiver. The bank, being a centralized authority, keeps records of these transactions, and only the two participating members and bank officials can verify the transaction. Blockchain, being a public ledger, works differently. No centralized authority is available to control transactions, and verification is made by designated participating members other than the members who initiated and received transactions [13].

1.5 BITCOIN AND BLOCKCHAIN

Many people erroneously assume that Blockchain and Bitcoin are similar. In reality, they are not similar, but they are functionally related. Put simply, Blockchain is a distributed database, which means data are duplicated among different nodes at different geographical locations. This approach of duplication leads to decentralization of data. Decentralization results in an enhanced level of security, as no central data repository is available to attack [15] (Figure 1.3).

Bitcoin is the first digital decentralized cryptocurrency that uses Blockchain network exclusively for its transactions. The users in this network transfer bitcoin and trade without any third-party intermediary. Currently there are many cryptocurrencies available in the global market, but Blockchain is the common platform for their execution.

1.6 BASICS OF BLOCKCHAIN TECHNOLOGY

A Blockchain platform is built mainly on two pillars, namely distributed database and data structure. There happens to be many interconnected nodes or workstations that initiate, validate transactions, store data, etc. Blocks in Blockchain form a list data structure having a cryptographic hash link between any two blocks. Each new block in the Blockchain incorporates some essential keys within itself [17]. The first block in a chain is known as a Genesis block or Block 0. It contains previous hash = 0. The newly created block is appended to the Genesis block, and any new

FIGURE 1.3   Bitcoin and Blockchain [16].

Hash	000000000019d6689c085ae165831e934ff763ae46a2a6c172b3f1b60a8ce26f 🗎
Confirmations	718,808
Timestamp	2009-01-03 23:45
Height	0
Miner	Unknown
Number of Transactions	1
Difficulty	1.00
Merkle root	4a5e1e4baab89f3a32518a88c31bc87f618f76673e2cc77ab2127b7afdeda33b
Version	0x1
Bits	486,004,799
Weight	1,140 WU
Size	285 bytes
Nonce	2,083,236,893
Transaction Volume	0.00000000 BTC

FIGURE 1.4   Records of block 0 in bitcoin Blockchain.

creation thereafter gets appended to the previous block. The first Bitcoin Genesis block was mined at 11:45 PM GMT on January 3, 2009; its creator has not been identified. Figure 1.4 shows self hash value, confirmation until date, Markle hash value, Nonce value, etc of the Genesis block of Bitcoin Blockchain.

1.6.1 New Block Creation Protocol

The flowchart in Figure 1.5 shows the creation of a new block. Line 1 keeps track of increasing length of chain. Line 2 sets the timestamp at the instant of formation of a new block. Line 3 incorporates proof. Line 4 shows inclusion of the previous hash value into the new block. Line 5 includes any transaction at the time of creation of the new block. Finally, the entire block is returned to the point of calling. Figure 1.6 shows the links between blocks. One block is linked to next block in the chain. A block has two sections in it, i.e., header and body [17]. Header contains some records like hash of previous block, hash of current block, hash of next block, Nonce value, height of the block in chain, miner information, transaction value, award information, time stamp, etc. Body contains hash of transactions. Once a block is added to the chain, nobody can alter anything within the block.

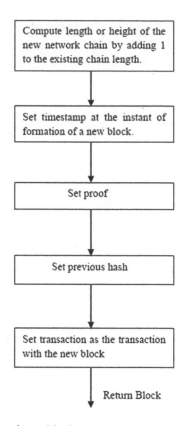

FIGURE 1.5   Flowchart of new block creation.

1.6.2 Hash Value Generation of the Newly Created Block

The flowchart in Figure 1.7 is a protocol for generating hash value of the newly created block. The algorithm in Figure 1.6 returns the block accepted by the HASH function of Figure 1.7. Line 1 encodes the accepted block. Line 2 converts the encoded value into hexadecimal hash number. This hexcode is returned.

1.6.3 Proof of Work

The flowchart in Figure 1.8 represents the Proof of Work consensus protocol. Line 1 sets parameter value Nonce to 1 as initial value. Line 2 initializes value of Check Proof to FALSE. The following lines compute the value of Nonce until first n digits of the hash value are all zeros. Finally return Nonce.

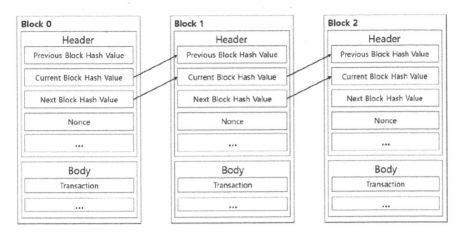

FIGURE 1.6   Blocks in Blockchain [18].

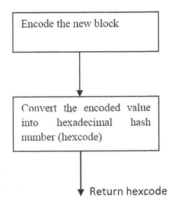

FIGURE 1.7   Protocol for generating Hash value of the newly created block.

1.6.4 Mining a Block in Blockchain

The flowchart in Figure 1.9 presents a mining consensus mechanism of a block in blockchain. This procedure adds a new block to the existing chain. The system determines a mining difficulty value. Records in the proposed block and previous blocks act as inputs to hash function and generate the hash value of the new block. This process continues until this hash matches with the target or difficulty set by the system. During each iteration the nonce value will be incremented by 1. Once this target is achieved, the puzzle is solved. Once the puzzle is solved, the mining process is completed and the proposed block will become the new block and will be added to the existing longest chain [19].

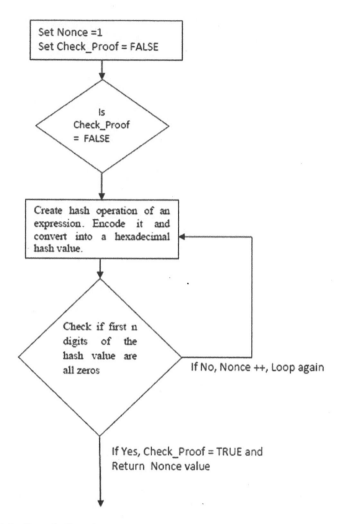

FIGURE 1.8   Proof of work.

1.7 MERKLE TREE

In Merkle tree, leaf nodes present data blocks, and non-leaf nodes are labeled with crypto-graphic hash function. Figure 1.10 shows a Merkle tree with data blocks and hash function. Data blocks at level 3 are taken as inputs to the hash functions of level 2. The output hashes at level 2 are further getting hashed at level 1, and final hash value is available at level 0, i.e., at the root of the tree. If we observe blocks at Figure 1.5, the body section consumes a lot of space, as the hash values of all transactions are stored there. A Merkle tree concept enables us to store only the Merkle

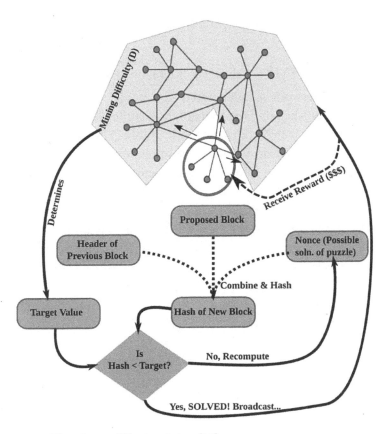

FIGURE 1.9   Flowchart of block mining [16].

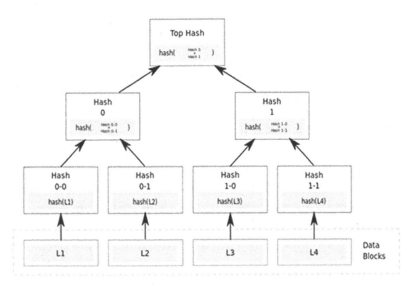

FIGURE 1.10   Merkle tree [20].

root in place of the number of transactions. This reduces storage space to a great extent [21].

1.8 DIFFERENT TYPES OF BLOCKCHAIN FRAMEWORK

A Blockchain network is categorized into public, private, and permissioned. Public Blockchain is open to all. It means anybody without revealing identity can participate in network. Public Blockchain works with incentive scheme, and on successful mining, a miner gets an incentive. The participants have to follow some consensus algorithms as discussed in the previous subsections. Public Blockchain consumes huge amounts of electrical energy and requires a huge amount of computational power.

Private Blockchain allows only selected participants through some authentication and verification mechanisms. This type has an owner and it will select participants and miner. The owner here can edit, alter, delete any entry inside a block as required.

The third type is permissioned Blockchain that borrows properties both from public and private types [22, 23]. It can allow anyone to join but must go through identification and verification procedures. The participants are allowed to perform some specific functions only.

1.9 BASIC ARCHITECTURE OF IoT

Basic or modern IoT architectures must have at least four different layers, namely the sensing layer, network layer, data-processing layer, and application layer. The sensing layer contains sensors and actuators within or outside physical objects. Sensors mainly collect data from environment and give them to actuators. Actuators conduct remedial actions in real time [24] (Figure 1.11).

The second layer is the network layer. It is mainly responsible for data transmission. This unit has a data acquisition system (DAS) and is directly connected to sensors. It collects raw analog data and converts them into digital form and the digital data into transmissible format. Finally, the data are transmitted through internet gateways for the next level of processing. In some cases the amount of data is huge. In those cases, data are filtered and compressed before transmission. The third layer performs data processing. This unit performs data analysis and data preprocessing before sending them to a cloud.

This is the last layer in the four-layer structure. In this layer cloud or data center further performs in-depth analysis, management, security,

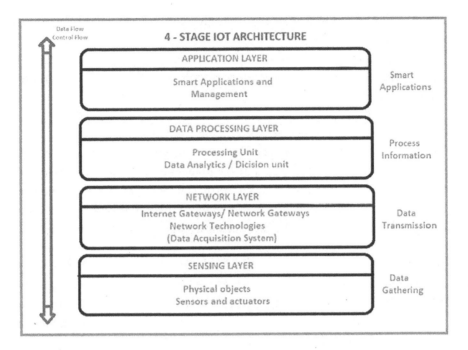

FIGURE 1.11   Basic architectural units of IoT systems [20].

and privacy of data. This layer is accessed by end-user individuals and organizations for different applications [25].

1.10 APPLICATIONS AND USE CASES OF IoT

According to the latest IoT Use Case Adoption Report 2021 [26], IoT adaptation rate is on the rise. This report also reflects some top applications of IoT, and this section addresses four of them. One of the widely used applications is smart home security. Some other applications such as home locking system, set-top box, and home surveillance system have evolved with the concept of smart home [27].

The next application of IoT is in smart city. Citizens have day-to-day challenges and concerns like waste management, traffic management, water management, etc. With application of IoT we may try to eliminate all those hazards [28].

Production of self-driven cars is the next application of IoT. Cars possess a number of sensors and devices to capture data from environment and send them to cloud through internet. They also make decisions using some machine learning algorithms. There is a challenge of delay in data circulation and decision-making, though a lot of researchis going on to address such issues [29].

IoT can address some concerns in other sectors, such as farming. IoT helps with drones for crop surveillance, understanding crop patterns and quality, irrigation, etc. [30].

1.11 DISADVANTAGES OF IoT ARCHITECTURES

We have seen a lot of advantages of implementing IoT in different real-life situations. Now we will talk about disadvantages of IoT. The main disadvantage is with security and privacy. We connect a lot of devices to the internet that may contain sensitive information. Hence there is always a chance of it being stolen or misused, as the network is a single point of failure [31].

1.12 IoT APPLICATIONS IN BLOCKCHAIN FRAMEWORK

In the present context, IoT solutions are based on centralized server communicating via internet. This huge architecture of IoT applications has to deal with a single point of failure. Hence an appropriate replacement is imperative, which may also provide enhanced security and transparency. Considering most of the demerits of a centralized system, here we are incorporating Blockchain as platform in different IoT applications. This proposal enhances security, stops any third-party interventions, and increases transparency.

Systems using IoT architectures face some security problems during data transmission. IoT architecture in a Blockchain framework can provide the required security and privacy to data. Integrating IoT applications in a Blockchain platform ensures data integrity, recording of immutable data, monitoring of real-time transactions, etc. Now let us talk about some basics of IoT and its architectures.

Internet of Things refers to physical objects communicating with each other through internet. There are a few fundamental characteristics like interconnectivity, heterogeneity, and dynamic changes that any IoT infrastructure possesses. With the property of interconnectivity, devices with different infrastructures can communicate. With the second property, communications are established even within different hardware, software, and network platforms. The third property means that the interconnected devices can dynamically switch their states, locations, speed etc.

Internet of Things (IoT) extends to industrial sectors as well, referred to as Industrial Internet of Things (IIoT). Here it encompasses communications between machines, unit devices, big data, machine learning, and other industrial applications.

Internet of Everything is another term often used today. According to CISCO, it is an intelligent communication between objects, data, people, etc. IoT mainly deals with physical objects and their interconnections. But the thin line between these two is the incorporation of intelligence to improve some existing drawbacks and making IoT systems more accomplished [32].

IoT projects on a Blockchain platform depend on two basic properties of Blockchain technology: distribution and immutability. With distribution property of Blockchain, database does not reside in a single location and is distributed throughout the nodes accessible by users. Immutability ensures originality of data in IoT projects with Blockchain. The next section reiterates the two IoT projects of the Blockchain platform.

1.12.1 Blockchain in Supply Chain Management

Supply chain management is the combination of upstream and downstream processes. The upstream supply chain process may include procurement of raw materials, management of logistics, and manufacturing of goods. The downstream supply chain process may incorporate finishing of the manufactured products and distributing them to the end customers [33]. Presently supply chain faces some challenges such as lack of transparency on product information, raw materials, origin of product, etc. It may also face problems with operating with an established database like ERP [34]. The third key issue with Blockchain in supply chain is legal. The users in Blockchain are located in different geographical areas. Each country has its own law, and there is no common law for Blockchain for its execution. Blockchain in supply chain brings back trust between untrusted participants. It also brings back transparency by enabling better visibility about product information, product origin, raw materials, logistics, etc. Any stakeholder can be a part of Blockchain network. Hence a common immutable information may be sent and recorded in real time to manufacturer, supplier, retailer, etc. (Figure 1.12).

```
Supply chain upstream processes
(Raw Material -> Suppliers -> Manufacturers)
```

```
Supply chain downstream processes
(Manufacturers -> Distributers -> Customers)
```

FIGURE 1.12 Supply chain management.

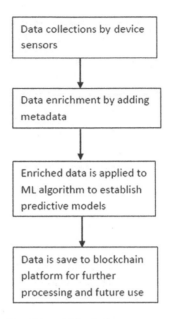

FIGURE 1.13   Flowchart for IoT and Blockchain in agriculture.

1.12.2 Blockchain in Agriculture

Intelligent IoT system on a Blockchain platform can help farmers in producing various crops. IoT sensors collect data like soil pH, moisture, humidity, temperature at different seasons, etc. These data are directly or indirectly related to the growth of crops. These data are to be stored in some structured format [35].

The data are to be enriched by adding some metadata like timestamp, demography, type, etc. This enriched data set is now ready for applying to machine learning algorithms. Machine learning algorithms establish some predictive models and recommendation system like quality prediction of crops, yield prediction, growth of crop, market demand prediction, etc. Finally data are saved in Blockchain storage for further processing and future use (Figure 1.13).

1.13 CONCLUSION

This chapter first discussed some general ideas and applications in Blockchain technology. It then discussed the history and evolution of the technology and various characteristics of Blockchain, particularly as a distributed platform. It then showed the difference between conventional and Blockchain-based transactions. It discussed the differences between Bitcoin cryptocurrency and Blockchain. It also discussed basics of Blockchain technology where it illustrated different consensus protocols like PoW, mining, and other basic parameters. Later it discussed basic architecture of IoT and its various applications, use cases, and disadvantages. Finally, it demonstrated the use of IoT applications in a Blockchain framework.

REFERENCES

[1] C. P. Fran Casinoa, T. K. Dasaklisb, "A systematic literature review of blockchain-based applications: Current status, classification and open issues," *Telematics and Informatics*, vol. 36, pp. 55–81, 2019.

[2] S. Cheng, B. Zeng, and Y. Z. Huang, "Research on application model of blockchain technology in distributed electricity market," *IOP Conference Series: Earth and Environmental Science*, vol. 93, no. 1, p. 012065, November2017. [Online]. https://doi.org/10.1088/1755-1315/93/1/012065

[3] R. Cheng, F. Zhang, J. Kos, W. He, N. Hynes, N. Johnson, A. Juels, A. Miller, andD. Song, "Ekiden: A platform for confidentiality-preserving, trustworthy, and performant smart contracts," in *2019 IEEE European Symposium on Security and Privacy (EuroS&P)*, June 2019. [Online]. https://doi.org/10.1109/EuroSP.2019.00023

[4] C.-W. Chiang, E. Betanzos, and S. Savage, "Exploring blockchain for trustful collaborations between immigrants and governments," in *Extended Abstracts of the 2018 CHI Conference on Human Factors in Computing Systems*, April 2018. [Online]. https://doi.org/10.1145/3170427.3188660

[5] N. C. K. Yiu, "Toward blockchain-enabled supply chain anti-counterfeiting and traceability," *Future Internet*, vol. 13, no. 4, 2021. [Online]. Available: https://www.mdpi.com/1999-5903/13/4/86

[6] E. Albert, P. Arenas, S. Genaim, and G. Puebla, "Closed-form upper bounds in static cost analysis," *Journal of Automated Reasoning*, vol. 46, pp. 161–203, 2010.

[7] I. Weber, X. Xu, R. Riveret, G. Governatori, A. Ponomarev, and J. Mendling, "Untrusted business process monitoring and execution using blockchain," in *International Conference on Business Process Management*, Springer, Cham, 2016.

[8] R. Dennis, G. Owenson, and B. Aziz, "A temporal blockchain: A formal analysis," in *2016 International Conference on Collaboration Technologies and Systems (CTS)*, vol. 102016, pp. 430–437.

[9] I. Weber, V. Gramoli, A. Ponomarev, M. Staples, R. Holz, A. B. Tran, and P. Rimba, "On availability for blockchain-based systems," in *2017 IEEE 36th Symposium on Reliable Distributed Systems (SRDS)*, 2017, pp. 64–73.

[10] Z. Duan, H. Mao, Z. Chen, X. Bai, K. Hu, and J.-P. Talpin, "Formal modeling and verification of blockchain system," in *Proceedings of the 10th International Conference on Computer Modeling and Simulation*, 2018.

[11] J.-H. Huh and K. Seo, "Blockchain-based mobile fingerprint verification and automatic log-in platform for future computing," *The Journal of Supercomputing*, vol. 75, no. 6, pp. 3123–3139, 2019.

[12] L. Ismail, H. Hameed, M. Alshamsi, M. Alhammadi, and N. Aldhanhani, "Towards a blockchain deployment at uaeuniversity: Performance evaluation and blockchain taxonomy," in *Proceedings of the 2019 International Conference on Blockchain Technology*, 2019.

[13] H. Sukhwani, N. Wang, K. S. Trivedi, and A. J. Rindos, "Performance modeling of hyperledger fabric (permissioned blockchain network)," in *2018 IEEE 17th International Symposium on Network Computing and Applications (NCA)*, pp. 1–8, 2018.

[14] "Blockchain tutorial: Learn blockchain technology," https://www.guru99.com/blockchain-tutorial.html, accessed: 2021-11-18.

[15] L. Aniello, E. Gaetani, F. Lombardi, A. Margheri, and V. Sassone, "A prototype evaluation of a tamper-resistant high performance blockchain-based transaction log for a distributed database," in *2017 13th European Dependable Computing Conference (EDCC)*, 2017.

[16] D. K. Tosh, S. Shetty, X. Liang, C. A. Kamhoua, K. A. Kwiat, and L. Njilla, "Security implications of blockchain cloud with analysis of block withholding attack," in *2017 17th IEEE/ACM International Symposium on Cluster, Cloud and Grid Computing (CCGRID)*, 2017, pp. 458–467.

[17] H. Sukhwani, J. M. Martnez, X. Chang, K. S. Trivedi, and A. Rindos, "Performance modeling of PBFT consensus process for permissioned blockchain network (hyperledgerfasbric)," in *2017 IEEE 36th Symposium on Reliable Distributed Systems (SRDS)*, 2017, pp. 253–255.

[18] J. H. Park and J. H. Park, "Blockchain security in cloud computing: Use cases, challenges, and solutions," *Symmetry*, vol. 9, no. 8, 2017. [Online]. Available: https://www.mdpi.com/2073-8994/9/8/164

[19] S. M. H. Bamakan, A. Motavali, and A. Babaei Bondarti, "A survey of block chain consensus algorithms performance evaluation criteria," *Expert Systems with Applications*, vol. 154, p. 113385, 2020. [Online]. Available: https://www.sciencedirect.com/science/article/pii/S0957417420302098

[20] "Architecture of internet of things (IoT)," https://wwsw.geeksforgeeks.org/architecture-of-internet-of-things-iot, accessed:2020-06-06

[21] S. Abeyratne and R. Monfared, "Blockchain ready manufacturing supply chain using distributed ledger," *International Journal of Research in Engineering and Technology*, vol. 5, no. 9, pp. 1–10, 2016.

[22] F. Gao, L. Zhu, M. Shen, K. Sharif, Z. Wan, and K. Ren, "A blockchain-based privacy-preserving payment mechanism for vehicle-to-grid networks," *IEEE Network*, vol. 32, no. 6, pp. 184–192, 2018.

[23] A. Hafid, A. S. Hafid, and M. Samih, "Scaling blockchains: A comprehensive survey," *IEEE Access*, vol. 8, pp. 125 244–125 262, 2020.

[24] S. Kumar, P. Tiwari, and M. Zymbler, "M. internet of things is a revolutionary approach for future technology enhancement: A review," *J Big Data*, vol. 6, no. 111, 2020.

[25] K. Gatsis and G. J. Pappas,"Wireless control for the IoT: Power, spectrum, and security challenges: Poster abstract," in *Proceedings of the Second International Conference on Internet-of-Things Design and Implementation*,

ser. IoTDI'17. New York: Association for Computing Machinery, 2017, p. 341342. [Online]. Available: https://doi.org/10.1145/3054977.3057313

[26] "IoT use case adoption report 2021," https://iot-analytics.com/product/iot-use-case-adoption-report-2021-2/, accessed: 2021.

[27] M. Luk, G. Mezzour, A. Perrig, and V. Gligor, "Minisec: A secure sensor network communication architecture," in *Proceedings of the 6th International Conference on Information Processing in Sensor Networks*, ser. IPSN'07. New York: Association for Computing Machinery, 2007, p. 479488. [Online]. https://doi.org/10.1145/1236360.1236421

[28] F. Behrendt, "Cycling the smart and sustainable city: Analyzing ec policy documents on internet of things, mobility and transport, and smart cities," *Sustainability*, vol. 11, no. 3, 2019. [Online]. Available:https://www.mdpi.com/2071-1050/11/3/763

[29] R. Patan, K. Suresh, and M. Babu, "Real-time smart traffic management system for smart cities by using internet of things and big data," in *2016 International Conference on Emerging Technological Trends (ICETT)*, vol.10, 2016, pp. 1–7.

[30] T. Qiu, H. Xiao, and P. Zhou, "Framework and case studies of intelligence monitoring platform in facility agriculture ecosystem," in *2013 Second International Conference on Agro-Geoinformatics (Agro-Geoinformatics)*, pp. 522–525, 2013.

[31] E. Park, A. P. Del Pobil, and S. J. Kwon, "The role of internet of things (IoT) in smart cities: Technology roadmap-oriented approaches," *Sustainability*, vol. 10, no. 5, 2018. [Online]. Available: https://www.mdpi.com/2071-1050/10/5/1388

[32] Z. Zheng, S. Xie, H.-N. Dai, X. Chen, and H. Wang, "An overview of blockchain technology: Architecture, consensus, and future trends," in *2017 IEEE International Congress on Big Data (BigData Congress)*, 2017.

[33] A. Banerjee, Blockchain technology: Supply chain insights from ERP, *Advances in Computers*, vol. 111, pp. 69–98, 2018.

[34] "Ibm blockchain, ibm global financing uses blockchain technology to quickly resolve financial disputes," https://www.ibm.com/blockchain/infographic/finance.html, accessed: February, 2018.

[35] H. Xiong, T. Dalhaus, P. Wang, and J. Huang, "Blockchain technology for agriculture: Applications and rationale," *Frontiers in Blockchain*, vol. 3, p. 7, 2020.

CHAPTER 2

A Survey on Various Applications of Internet of Things on Blockchain Platform

Archana Yengkhom and Debarka Mukhopadhyay

Christ University, Bangalore, India

CONTENTS

DOI: 10.1201/9781003188247-2

2.1 INTRODUCTION

As of 2021, the world required innovative solutions to tackle the fast-growing economy and its insatiable demand for technology that will satisfy the society's needs in terms of security, privacy, speed, functionalities, data management, and much more. This was evident during the COVID-19 pandemic that continues to drive the economy and its people to perform and deliver better to fight against time and money. From the front-line industries such as the healthcare and supply chain, at times struggling to deliver the most needed resources like oxygen and life-saving medicines across states and countries, the need for more innovative solutions could not be underappreciated. With the growing need for better understanding and study of the systems that could help catapult the existing systems to exponentially increase its performance, Blockchain technology (BCT) for the Internet of Things (IoT) represents an innovative solution that has proven to play a revolutionary role in evolving the existing system in various sectors (healthcare [1], supply chain [2], smart cities and homes [3, 4]). With data playing the role of a shadow currency for the current economy, BCT cannot be overemphasized in its key role as a game changer.

BCT has captivated the world ever since its first introduction by Satoshi Nakamoto in 2008, mainly due to its immutable distributed digital ledger, for a decentralized trust that provided the world a moment of escape for the economy to be run without the interference of any central authority, government, or requirement for any Trusted Third Parties (TTP). Although BCT has its fair share of the spotlight in the market, there are many other problems yet to be discussed: the cumbersome load of large transactions, possible attacks [5], the limit on the size of chains that could be handled, the necessity of miners, and its inevitable need for transaction fees. Nevertheless, BCT has empowered and continues to drive many researchers and businesses with its benefits in terms of security, scalability, privacy, transparency, data integrity, management, and traceability.

On the other end, we also saw the emergence of the Internet of Things (IoT) or Internet of Everything, its high demand during the last few decades, and its remarkable achievements. IoT, a technology that drives on distributed connectivity of devices, could not resist the inquisitive minds of many that saw the potential and relevant benefits of the integration of BCT and IOT; it presented a solution for overcoming the challenges posed by IoT applications in terms of its security and scalability fueled by its centralized framework, Single Point of Failure (SPF), many-to-one nature of connectivity, and protection of privacy [2, 6]. At the end of 2018, it was estimated that 22 billion devices were connected, and by 2025, the number is expected to jump to 38.6 billion [6]. According to *Economic Times*, it is anticipated that at least 10 devices will be connected to I0T in every household of a developed country, and possibly even more with the integration of 5G technology [7]. This data demonstrate that the market of the IoT paradigm, one of the most recognized future technologies, is yet to reach its peak as illustrated in the Gartner Hype Cycle,2017 [8]. Thus, the urgent need for more collaborative efforts, experimental works, and global attention is presented in the following sections to overcome the challenges of IoT applications to bridge the gap from theory-based to more real-world applications.

Over the last few years, like-minded enthusiasts and academicians have been working on overcoming the challenges and limits demonstrated by BCT and IoT by adopting new methodologies, approaches, and alternatives. Some of these studies are discussed in the following sections to corroborate the multitude of research conducted on Blockchain-based IoT environment: Fabric-IoT [9], Blockchain Mirror Model [10], index Access Management in IoT Access Management in IoT [6], Blockchain Signaling System [11], and Lightweight Scalable Blockchain for IoT (LSB) [4] and Hawk [12].

The chapter addresses the current trends of applications based on Blockchain for IoT and case studies of real-world applications deployed or under development in the market. The study aims to explore and gather the evidence based on recent research conducted and its capability to deliver to the market. Healthcare sector and supply chain management are among the most sought-out industries, as they play a crucial role in the economy. They will also be the most beneficial recipients, as they will dominate most of the market according to current trajectory of the economy. With advancements of human lifestyles and quality of life being an indicator to a nation's development, the need for more convenient, transparent, secured, and real-time data connectivity becomes globally accepted, and the demand for applications in most industries (like e-commerce, agriculture, public safety, smart homes, and smart cities) will also parallel the need to fulfill the consumers' and stakeholders' quality demands without a compromise on data integrity and security.

Alternatives to Blockchain, such as DAG, are discussed to show that Blockchain is unlikely to be the only viable solution. With various challenges and risks involved with the technologies, the study also highlights alternatives and concludes why BCT for IoT is among the most viable options. This hope represents a concise understanding of current and future trends of the market, economically and socially, and draws a picture of the relevant information required by the reader.

2.2 OVERVIEW OF BLOCKCHAIN TECHNOLOGY

Blockchain technology (BCT) saw potential disruptions to the existing system as the applications and transactions, which required trusted third parties or a centralized system, could now operate securely in a distributed decentralized peer-to-peer (P2P) network. The fundamental basis for Blockchain architecture offers key features such as traceability, transparency, security, integrity, and trust in the system. BCT paved a new path and standard for other technologies to compete against and meet those requirements to satisfy their stakeholders. It works as a P2P-enabled network to communicate using cryptographic algorithms. With no centralized node, Blockchain technology enables users to store data in a decentralized system using distributed ledgers. Bitcoin, Ethereum, Stellar, Hyperledger, etc. are some of the popular Blockchain implementations that provide an insight to the potential of its disruptive technology.

2.2.1 Architecture

Current popular databases are based on a client/server architecture, a centralized access control system. The client plays the role of receiving nodes, and the server maintains the central authority in a distributed system. This was shaped to meet the traditional human–machine interaction.

Meanwhile, Blockchain is founded on a decentralized system, where each node in the network communicates on a P2P protocol. Blockchain technology, as the name suggests, is a sequence-chain of blocks that store transaction records, digitally and tamper-proof. This is achieved via the concept of a Merkle Tree or Hash Tree; it is a tree-based data structure that maintains data integrity and synchronization for a distributed network by exploiting the hash functions. A cryptography hash (digest) acts as a digital signature due to its uniqueness in terms of text or data. A hash function is a method that ensures the immutability of data as it maps a unique output for a specific input; the input cannot be derived from its output. This is further enhanced by using hexadecimal for maturing the SHA-256 bit (Secured Hash Algorithm) cryptography function. It produces a unique signature and plays a key role in the data integrity for BCT.

A typical Blockchain structure is illustrated in Figure 2.1. The initial block is referred to as the Genesis block. Each block comprises two parts: a block header and a block body. The block header will store information about the block; it stores the block version, the root hash, a time stamp, n-Bits, hash of the previous block, and Nonce value that is added or encrypted with the hash value to add another layer of security. The block body stores information about the transactions. A transaction in Blockchain refers to an event triggered by the participant. The transaction data stores the root hash of the Merkle Tree. Each transaction is denoted as T_x as illustrated in [13]. When a new transaction is appended to the existing chain, validation of the transaction is first computed and competed among the participants (miners) by following an accepted

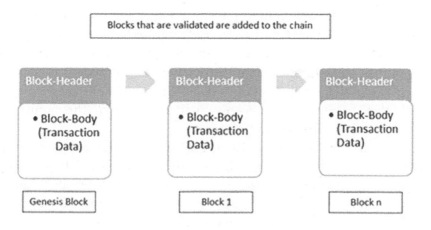

FIGURE 2.1   Architecture of a Blockchain model.

consensus algorithm to maintain the integrity of the chain. Forks are multiple chains created at close moments; this is resolved by selecting the latest block for the longest valid chain. The blocks in the shorter chains are referred to as orphan blocks.

2.2.2 Consensus Algorithm

The consensus algorithm is inspired by the Byzantine General Problem (specifically for Bitcoin) [13], which considered an army commander preparing to launch an attack on a large enemy base. As the area to be covered was very large, the general was faced with a dilemma of informing the entirety of his army about the plan to attack without compromising the secrecy of the order. The reason for his hesitancy was the fact that the message could be intercepted by his enemy or one of his own soldiers could betray him by leaking the message. A consensus for his action had to be achieved. In such a distributed environment, it raised many challenges that could similarly be mapped with the Blockchain distributed network.

The following are the most commonly adopted solutions mentioned in [14]:

PoW (Proof of Work): As described in detail in [15], a PoW is achieved by solving a puzzle and calculating its value. All the participant nodes compete to calculate this unique hash value within a time period. The nonce located at the block header of the latest block is frequently changed so when a node computes the desired value and claims first, it is validated by all the other participants. Once verified, a new block is coined to be 'mined' and added to the chain.

PoS(Proof of Stake): From [14], this is another approach to PoW. PoS was introduced to further the cause of consensus as a more reliable approach. In the case of PoS, the participant nodes must prove the ownership of their stake (proportion of currency in use). A node is randomly selected as a validator in terms of its stake. The most important feature of PoS is that it consumes less energy compared to PoW. This is further explained in [14] to overcome its fairness.

PBFT (Practical Byzantine Fault Tolerance): Introduced to tolerate Byzantine faults. It can only handle one-third of the replicas. It mandates the publicity of all the participant nodes in the system. Hyperledger Fabric takes advantage of this algorithm. It requires validation from at least two-thirds of all the nodes to enter a new phase (Three phases: pre-prepared, prepared, and commit).

DPoS (Delegated Proof of Stake): Mentioned in [14], and as the name suggests, DPoS is a result of a democratic policy where stakeholders in the network nominate their delegates to act as validators and miners. This significantly reduces the number of participants and provides a contingency plan for any selfish approaches, as the delegates can be removed at any time.

Ripple: A recent development [14], it shifts to a subsystem of trusted nodes from a general network. Nodes that are involved in the consensus process are termed *servers*, and the clients are the nodes involved in transacting funds. The servers maintain their own unique node list (UNL) to keep a check by mandating at least 80% approval from the UNL.

2.2.3 Digital Signature

A digital signature is like a unique fingerprint that is used to identify the credentials of a user. This is achieved by cryptography functions. The most commonly used algorithm in the Blockchain is the elliptic curve digital signature algorithm (ECDSA) mentioned in [16]. ECDSA creates key pairs for the signing and verification phase of a digitally signed transaction.

2.2.4 Mining

Miners is the term used for participants in the network that compete to calculate the hash values in Blockchain, and the process is called *mining*. Analysis of mining plays a key role in incentivizing the dynamic of a Blockchain model. Various research is conducted in game theory–based solutions to better analyze and predict mining behaviors and provide optimal reaction strategies. As per definition, game theory is a study of mathematical models of strategic interaction between rational decision-makers. It is used to analyze the various games strategies of the consensus nodes and their interactions to optimize the mining (Nash's Equilibrium, etc.). It can also be utilized to develop incentive mechanisms that discourage the nodes from executing misbehavior or launching attacks.

2.2.5 Smart Contracts

Introduced by Szabo [17], a smart contract is a digital contract that gets bound within a system. They are automatically executed when the conditions are met, reducing time and the need for a mediator. Hence, smart contracts play a big role in verifying and enforcing conditions in a Blockchain as required. Proposals and various other approaches have been identified in [18] (Figure 2.2).

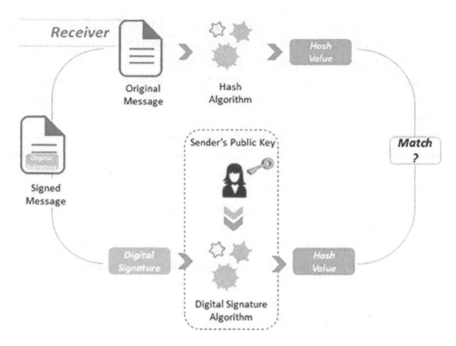

FIGURE 2.2   Digital signature.

2.3 OVERVIEW OF IoT TECHNOLOGY

Internet of Things (IoT) has been growing at an incredible rate in terms of its achievements and global acceptance because of low-power lossy networks (LLNs). It is a distributed network of things (devices, sensors, etc.) within a public/private network. The interconnectivity of devices as sensors and actuators bridged the gap of sharing information, and communication between the devices provided new and exciting opportunities for enterprises and the world to reap its benefits. IoT is the new Internet for embedded devices. Its application varies from homes, cities, farming/agriculture, energy systems, health and fitness monitoring to smart payments, smart grids to weather monitoring to IoT-based supply chain solutions for most of the industries, improving the efficiency and performance from current systems (which rely heavily on manual settings and configuration of devices involved). Machine-to-machine and machine-to-human interactions formed the backbone of the motivation behind IoT. The scope of IoT is not limited to just the connectivity of things; it can communicate and exchange data while implementing value-added application-based services. Applications on IoT networks extract and create information from lower data by performing certain processes like filtering, categorizing, condensing, and contextualizing the data. This information is structured to interpret the knowledge toward its intended objectives, allowing for a smarter performance.

The 2015 paper by In Lee et al. [19] described the five popular technologies—RFID, WSN, Cloud, middleware, and the application software—required for deployment of IoT-based products and services. These IoT technologies were examined to determine the approaches that can be justified for IoT investments. Extensive research is being conducted to identify the various challenges for its deployment in a real-world application. Many enterprises, from IBM and Coca-Cola to Rio Operations Center (to name a few), have already invested and are successfully deploying IoT-based solutions.

2.3.1 IoT Architecture

An IoT device consists of several interfaces for connections (wired and wireless) such as Input and Output interfaces for sensors (UART, SPI, 12C, CAN), memory and backup interfaces (NAND/NOR, SD, MMC, etc.), interfaces for internet connectivity (USB Host, Ethernet), and multimedia interfaces (HDMI, RCA video, 3. 5mm audio, GPU). It can collect a wide range of data types over different systems, communicating over each other or cloud-based storage. IoT protocols mentioned in [20] illustrates the four layers as:

1. Link Layer: Some relevant protocols for the layer are the 802.3 (Ethernet), 802.11 (Wifi), 802.16 (WiMax), 802.15.4 (Low-rate wireless personal area network (LR-WPAN)), and 2G/3G/4G/5G mobile communication.

2. Network/Internet Layer: The network handles host addressing and packet routing. It sends IP data grams (contains the source and destination addresses) from the source network to the destination network across multiple networks. Hosts are identified using IP addressing schemes of IPV4/IPV6/6LoWPAN.

3. Transport Layer: The transport layer enables end-to-end messaging transfer independent of the underlying network. This is achieved by either using handshakes (TCP connection-oriented) or acknowledgments (UDP, a connection-less protocol). It is responsible for error, flow, and congestion control.

4. Application Layer: This layer defines how the application interface communicates with the lower-level protocols. Port numbers

are referenced for application addressing like port 80 for HTTP, port 22 for SSH, etc. HTTP (Hypertext Transfer Protocol), CoAP (Constrained Application Protocol), Web Socket, MQTT (Message Queue Telemetry Transport), XMPP (Extensible Messaging and Presence Protocol), DDS (Data Distribution Service), and AMQP (Advanced Message Queuing Protocol) are some of the widely used protocols in the application layer.

2.3.2 Logical Design of IoT

This section refers to the abstract representation of the devices and processes without detailed low-level specifications. The commonly used IoT functional blocks provide the system the ability to identify, sense, actuate, communicate, apply, secure, and manage concerning its environment and implementation as described in [20]. Some of the fundamental communication models that drive the IoT model are:

- Request-Response: Here the client sends requests to the server and the server determines the request and responds accordingly.

- Publish-Subscribe: It involves publishers, consumers, and a mediator known as a broker. Consumers subscribe to the topics managed by the brokers who in turn forward it to a publisher.

- Push-Pull: In this model, the data is pushed to queues and the consumers pull the data on a LIFO (Last in First Out) basis. Producers are unaware of the consumers, and the queues act as a buffer between the push and pull rates.

- Exclusive Pair: It is a bidirectional, fully duplex communication model between the client and the server. It is a form of an open connection.

2.3.3 IoT-Enabling Technologies

As mentioned previously, IoT is enabled by several core technologies, such as wireless sensor network (WSN), which is a collection of distributed devices with sensors that can monitor environmental conditions. It contains several end-nodes, a router, and a coordinator. The coordinator collects the data from all the nodes, behaving as a gateway that connects WSN to the Internet. WSN works with IEEE 802.15.4 (Zig-Bee), one of the most popular wireless technology. It operates at 2.4-GHz frequency and ranges from 10 to 100 meters. Several cloud computing technologies are also employed in IoT, such as Infrastructure as a Service (IaaS), Platform as a Service (PaaS) and Software as a Service (SaaS). Another important feature of IoT is the ability to analyze business and big data. As big data is generated by sensors and devices, the capability to discover and make analyses to make a real-time decision is a testament to transforming IoT-based solutions into value-added products and services.

2.3.4 IoT Levels

To truly understand the functioning of an IoT system, it is important to highlight the various levels of IoT systems. An IoT system consists of components such as a device, the resources (software components), a controller and web service components (Web Socket protocols), the database (either local or cloud-based), an analysis component, and finally the application as an interface to interact between the users and devices. Illustrated in [20], there are six IoT levels. Each level improves accordingly with the inclusion of cloud-based analytic tools introduced from level 3. Levels 4 and 5 have multiple end nodes and can handle more complex data. Level 6 improves on these further by working with independent multiple-end codes using a centralized controller, a cloud-based application for visualizing the data.

We have so far laid out the technical aspects of both Blockchain technology and IoT systems individually. In this section, we highlight the better known existing systems and compare them to both technologies. The goal is to understand the benefits and highlight the true potential of the mentioned technologies in the current economy. The traditional database has evolved tremendously over the last few decades. From a typical file-processing system to the current relational database management system (RDBMS) and further to even cloud-based (AWS, Azure, Oracle, etc.), it is difficult to overstate the role the development of its functionality has played and continues to play in the current economy. With data often used as the shadow currency of today's economy, the management and security over these data require better and better solutions. As highlighted in [21], the critical analytic comparison of a traditional database system and Blockchain is highlighted. Source [6] mentions that time plays an important role in database systems (Table 2.1). Database systems model the real world. With more and more data being accumulated, and the need for more transparent, secured, and decentralized systems to play the winning game, academicians and businesses across numerous industries opt for a more innovative solution. Blockchain technology fulfills most of that

TABLE 2.1   A Comparison Table of Existing and Proposed Blockchain Technology [22]

Issue	Blockchain Based	Central Database
Trust	Does not require a trusted party	Require a central trusted party
Confidentiality	All nodes have visibility of data	Access restriction
Fault Tolerance	Data is shared among the nodes	Data is stored in a central storage
Performance	Time lag in achieving consensus	Immediate execution
Redundancy	Each party has the latest copy	Copy is stored only at the central database
Security	Implements cryptography measures (Hash)	Implements traditional access control

requirement as seen in [22]. Now considering IoT, the technology has almost become a lifestyle for many developed nations. With an ecosystem that does not seem to cease growing anytime in the future, this only invokes many to wonder what the current economy would even be without IoT. From the concept of connected devices to actual wireless network applications since the 1990s, the need for connected devices to build a smart system has become a synonym to a powerhouse of a nation's technological advancement. From the vast research that mentions the benefits and capabilities brought by IoT [4], a traditional view of devices is almost obsolete. However, as with many new technological innovations, the issues and challenges that follow must be addressed and countered appropriately to create an efficient and better global standard. The challenges mentioned above under the IoT overview highlight some of the most important problems, especially in terms of security and privacy of data. Figure 2.3 illustrates the developments since 2010 and future applications [23]. By 2020, IoT-based applications dominate most of the industries from tracking and monitoring to dynamic self-adapting systems. The future trends predict a system of higher computations that would be more convenient and add value to its services and products.

2.4 OPPORTUNITIES AND THEIR SOCIOECONOMIC IMPLICATIONS

In the preceding sections we discussed the various aspects that build on Blockchain and IoT and its corresponding issues that challenge its scalability and performance. Further research and studies by various academicians and enthusiasts have correlated the architectural design elements of IoT with Blockchain technology that show promising results in bridging the gap of the limitations faced by IoT and enhancing its features.

IoT Roadmap from 2010 to 2050 and beyond

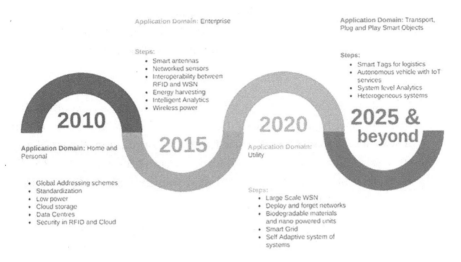

Application Domain: Enterprise

Steps:
- Smart antennas
- Networked sensors
- Interoperability between RFID and WSN
- Energy harvesting
- Intelligent Analytics
- Wireless power

Application Domain: Transport, Plug and Play Smart Objects

Steps:
- Smart Tags for logistics
- Autonomous vehicle with IoT services
- System level Analytics
- Heterogeneous systems

2010

2015

2020

2025 & beyond

Application Domain: Home and Personal
- Global Addressing schemes
- Standardization
- Low power
- Cloud storage
- Data Centres
- Security in RFID and Cloud

Application Domain: Utility

Steps:
- Large Scale WSN
- Deploy and forget networks
- Biodegradable materials and nano powered units
- Smart Grid
- Self Adaptive system of systems

FIGURE 2.3   The road map described in [23] for IoT technology.

The technical aspects that build on Blockchain, such as decentralized systems, digital signatures, and cryptographic consensus algorithms, play an important role for IoT. This is elaborated in [23]: the privacy feature of Blockchain where transactions are protected can be exploited by IoT applications to provide anonymity; the mining principle can be leveraged in IoT applications to incentivize the community and stakeholders for resources in domains like smart cities and supply chains, etc., shifting from a centralized system to a decentralized network in IoT for applications like supply chain like monitoring, transactions, tracking, etc. The most important advantage is the use of smart contracts in IoT applications for security and data authentication. Blockchain-based IoT has the potential to offset each other's technological limitations, such as Blockchain's proof of work (PoW), which may prove to be detrimental for an IoT solution because it requires high computational energy and power. The introduction of new innovative solutions like lightning network (LN) can address these limits and allow stakeholders to focus on its beneficial results [19].

2.4.1 Socioeconomic Impact

It is also important to consider the socioeconomic impact that will reverberate with society and the world after the implementation of Blockchain-based

IoT solutions. This is important, as it will form the backbone of the success for any product of innovative/inventive solutions. From ethical and moral relations to the growth of a nation's development, socioeconomical values need to be considered. With Blockchain-based IoT solutions, from research conducted over the past few years, this section concludes some of the important factors that will determine its global acceptance beyond its technological benefits.

1. Decentralized System: With a fast-paced economy, it becomes difficult to distinguish a trusted party from a potentially harmful investment. Trust is a very loose term yet a fundamental part of any successful business, so the need for a system without any intermediary interception has become an ideal solution, especially within a distributed network.

2. Connectivity: Communication has now expanded its boundary from just beyond human-to-human to machine-to-machine (M2M) to human-to-machine interactions with incredible efficiency. The challenges offered by IoT with Blockchain will drive many stakeholders to invest more in coming days.

3. Security and Privacy: Despite Blockchain's and IoT's limitations mentioned in [8], as with many new technologies, the collaboration of the two technologies has the potential to be the next deliverable solution for the current societal needs in terms of privacy and security it offers. This is due to the cryptography elements of a Blockchain and its ability to be anonymous. With IoT's vast network of connected devices, the vulnerabilities of IoT-based solutions can be overcome with the features provided by a Blockchain.

4. Data and Information: The current economy is run by a shadow currency called data. Thus, from big data to data management, Blockchain-based IoT solutions present themselves as one of the most fitting advantages compared to many other technologies, as they allow interoperability of multiple technologies, sharing of information with ease, and replication of data in multiple nodes ensuring its authenticity and integrity.

5. Low Cost: As both Blockchain and IoT work on existing systems such as devices, low-cost sensors, RFID tags, existing network, and

software protocols, the investment cost for a business setup will be reasonable and even lower than current ones in certain applications.

6. Quality of Life: The solutions stemming from Blockchain-based IoT technology will drastically improve the lifestyle of any society, as evident from their implementations. Many applications will help either solve problems or improve the convenience of our day-to-day work.

7. Applications: The opportunities that come with these technologies have already been mentioned in the earlier sections of this chapter. Application ranges from finance, health, and retail to supply chain and more, breaking the boundaries of its limitations provided from its contributions.

2.4.2 Emerging Technologies Relating to Blockchain with IoT

This section covers several emerging technologies associated with Blockchain-based IoT solutions to understand its current work and research conducted in this field. It will also highlight the solutions proposed to overcome the challenges presented by the technology. Blockchain technology, a system built to enable Bitcoin, the first cryptocurrency based on limited resources and constraints, has evolved over the years with a multitude of alternatives and breakthroughs, such as Ethereum, proposed in 2013 by Vitaliy Buterin. Ethereum is the backbone of modern smart contracts [17], which have applications beyond its initial intended functions. These smart contracts can help manage the IoT devices in terms of energy-saving features for appliances that are connected through the system [14]. IOTA [24], another breakthrough technology based on directed acyclic graphs, is built on Blockchain and IoT for the working of a distributed digital ledger. This is further elaborated in the later subsections.

2.4.2.1 Fabric-IoT

An access control system based on Blockchain for IoT application-a Fabric-IoT [9] works on the Hyper ledger Fabric framework and attributed-based access control (ABAC). It introduces three smart contracts, namely access contract, device contract, and policy contract. It aims to address the limitations posed by IoT in a dynamic scenario like mobility, limited computability, vulnerability access, etc. The paper demonstrated its results based on simulations conducted. This was met with desired results that could potentially help overcome challenges faced by IoT.

2.4.2.2 Blockchain Mirror Model

The researchers in [25] aim to enhance the data dependency of an IoT system by setting some preconditions to facilitate trust in a Blockchain. As the data that are produced or retrieved from these IoT devices lack a firm secured channel, the data may not always be accurate or may contain the breach. The mirror model functions by representing the real world and mapping it onto a trusted model referred to as a trust domain. The objects that model the real world are required to be time-stamped to ensure their integrity, and the transformative process of the objects needs to be synchronized to avoid any discrepancies. A smart contract is deployed to maintain its transactions with any new mirror assets. The paper showed promising results in various applications, one mentioned for biomedical engineering where the data collected requires appropriate data in response to the patient's needs. Figure 2.4 depicts the synchronization mechanism achieved through an API [25]. Most IoT devices have limited resources in terms of computing power, memory and storage capacity, and energy restrictions, while Blockchain network requires maintaining a copy of the data in every participating node to track the transactions. There are many proposed lightweight solutions, but they are not yet the most efficient. Researchers in [6, 9] proposed a management hub (like a

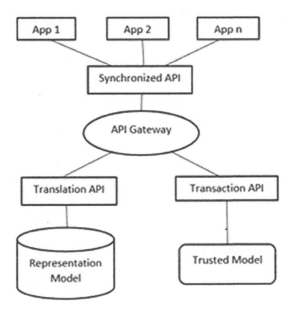

FIGURE 2.4 Synchronized API and API gateway.

miner node) that translates encrypted CoAP messages into JSON-RPC messages to communicate with the Blockchain. This process involves setting up the network, registering the device to the management hub, management protocols, and policy protocols to secure and manage access of the IoT devices onto the Blockchain network. The versatility of the introduction of management hubs demonstrated the capability of the system to scale and adapt to different scenarios. Security analysis also shows that the role of a manager in the network helps in avoiding spoofing and tampering possibilities.

2.4.2.3 Blockchain Signaling System (BLOSS)

The Blockchain signaling system (BloSS) [26] is a solution for security management, designed for cooperative defenses (like distributed denial of service [DDOS] attacks) by using software denied networks (SDN). BloSS was designed and prototyped in a global experiment to build specific protocols and registries to overcome the attacks and improve security vulnerabilities for a distributed network. This is built on Blockchain and could be further exploited in hybrid systems of integrated Blockchain-based IoT systems. Figure 2.5 illustrates the architecture design of BloSS [26], where we can observe the networking infrastructure being built on a decentralized storage structure using Ethereum.

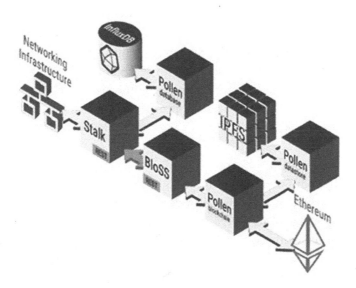

FIGURE 2.5   Architecture of BloSS [26].

2.4.2.4 Lightweight Scalable Blockchain for IoT

Many researchers have been working on developing a lightweight network to improve the features of many small devices based on their limitations of low computation and energy resources. This motivated IoT-based solutions that could be game-changing if achieved. Lightweight scalable blockchain (LSB) mentioned in [4] introduces a distributed time-based consensus (DTC) algorithm that will minimize the Blockchain's mining process. It works with multiple clusters to verify new blocks in the chain for transactions in the IoT system for overhead delays. At the same time, the paper demonstrates qualitatively that LSB has strong security performance in terms of attacks that were illustrated.

2.4.2.5 Hawk

As described in [12], Hawk is another solution for a decentralized smart contract system that can retain transaction privacy using an efficient cryptography protocol. The model suggests an improvement based on security for a distributed ledger, Blockchain. The user needs zero knowledge of cryptography, as the proposed compiler will automatically follow its encryption protocols. This can potentially be used for future IoT-based solutions, where setting cryptography elements of the Blockchain-based IoT solutions would be avoided, which would help in maximizing its performance while integrating with other technologies.

2.4.2.6 Hybrid System Based on Blockchain, IoT, and AI

With artificial intelligence as the tool for analytics, the Blockchain-based IoT system can further be enhanced in its applications to achieve a more efficient solution as suggested in [10]. Figure 2.6 illustrates the use of a cognitive process layer to analyze the data and optimize its management. The merging of AI, Blockchain, and IoT provides a new and exciting interest to many researchers.

2.4.2.7 Health Chain

The Health Chain mentioned in [25] is a healthcare industry application based on Blockchain—an IBM Foundation initiative. It integrates IoT and cloud computing to achieve a transformative result of the Blockchain. It provides authenticity, security, and data protection for a health sector automatically, thus resulting in faster and more confidential data management.

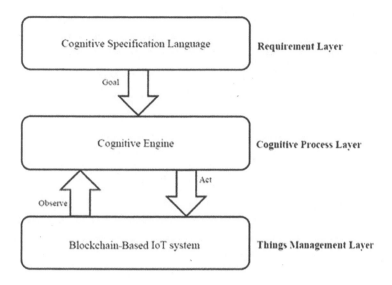

FIGURE 2.6   Architecture of cognitive-based layer mentioned in [10].

2.4.3 Applications

Blockchain-based IoT applications are gaining momentum in various sectors from supply chain to health to industrial sectors like manufacturing and management. This has prompted many enterprises and investors to focus on improving their operational performance. From IBM to KPMG and many others, applications explored are further described below.

2.4.3.1 Use Case-Based Applications

1. Supply Chain: One of the biggest sectors that would benefit from these solutions is the supply chain, as IoT can help in keeping track of various transported and stored goods from vehicles to warehouses using RFID tags, sensors to monitor their conditions, and devices to connect and communicate in real time. Blockchain plays an important role in managing and tracking the transactions and securing the data to ensure their integrity and authentication. Smart contracts also play a major role in complying with the standards of the goods and policies between the stakeholders.

2. Tracking and Monitoring of Business: Blockchain-based IoT applications are useful in monitoring and tracking the business in terms of maintaining the standard of goods and services offered by the

stakeholders. The paperwork is reduced drastically while maintaining its transparency, in the process ensuring trust and validity in the system. Everything is tracked in real time and automatically in a shared network setting with the required standard and compliance from all members. Its immutability covers most of the trust system and encourages individuals to follow up and maintain their integrity over their deliverables.

3. Smart City and Smart Homes: IoT enthusiasts love to talk about smart cities and smart homes, but due to its security and privacy challenges, the scalability and acceptance by society are restricted, inhibiting the potential that was proposed. The integration of technologies like Blockchain enables its application to services like home security systems, e-governance, auto-payments, monitoring, energy-saving applications, and data communication over realtime. Smart homes can offer synchronization of devices like updating stocks in a refrigerator by performing transaction-based decisions through a trusted and secured system. Data privacy of the homeowners can also be managed. Opportunities presented are very interesting for many enterprises and for urban sectors.

4. Data Management and Analytics: Blockchain-based IoT solutions have shown potential in enabling various predictive analyses for business and finance policies, services, and products. These include consultancy work by analyzing the pattern uncovered from multiple sources while supporting the integrity of the data. With big data, the need for data management and access controls is important and determines many factors in the decision-making processes. Businesses run on the integrity of the data and transaction processes, which the Blockchain offers. IoT plays the role of collecting data across various devices and locations with real-time data, thus improving its efficiency and reducing the loss.

5. Smart Health Care: The healthcare industry has several use-case-based applications for a Blockchain-based IoT solution. From tracking and monitoring of medicinal goods and pharmaceutical industries to maintaining its standards and policies using smart contracts, the applications can be potentially lifesaving. As the integrity and compliance with its regulations cannot be compromised, smart contracts and Blockchain can offer the support required for an IoT system.

6. Smart Farming: Farmers can now utilize the immutability and transparency offered by Blockchain-based solutions to not just eliminate the middleman but provide a clear and convenient way of boosting the work process. From monitoring the soil content to watering to controlling the temperature and humidity, these processes can all be automated without compromising the standard and quality of goods. Even transporting, selling, and buying goods can be done with confidence and trust in the system. As Blockchain provides immutability and traceability, end-to-end synchronization between the consumers and source can be validated and maintained.

7. Resource Management: Resource management has become an essential sector with the deeper realization of the finiteness of the world's resources. From water management to energy management, Blockchain-based IoT solutions can help solve the waste crisis. Wastage of resources left unmanaged can be drastic, and Blockchain-based IoT solutions can help reduce the energy footprint using its transparency and smart contract features to regulate the integrity of the commitments made by the stakeholders and the society.

8. Manufacturing and Automotive Industry: As production line demands high operational efficiency and quality checks, most of the automotive and manufacturing industries have been deploying sensors to regulate and monitor their activities. In recent years, many companies and enterprises have turned to IoT-based solutions to match the demand. With the integration of Blockchain, the transactions between various activities, stakeholders, and devices within the system can be secured and provide a level of regulation that is ideal in an industry. This can be seen in smart parking, fuel payment process, maintenance of parts and pieces of machinery, etc.

2.4.3.2 Real-World Applications

Looking at some companies and organizations that are currently adopting this technology for their application-based solutions leads to a reasonable argument in favor of the potential benefits that can turn Blockchain-based IoT applications into a trillion-dollar industry. Financial institutions, like ING, Deutsche Bank, and others, and supply chain giants like KPMG have been already adopting these technologies into their business model.

Golden State Foods (GSF), a known chain supplier in manufacturing and distribution of food products serving thousands of their clients, are currently working with IBM to scale and produce high-quality products using Blockchain and IoT. Devices like sensors collect data and secure it through the Blockchain for accountability and transparency.

Telstra is an Australian media company investing in smart home services. They have implemented Blockchain-based biometric security to capture the data using their devices, as this provides security over their data.

Mediledger, a pharmaceutical-based company, is employing Blockchain with IoT to track any legal change of any prescribed or owned medicines, which helps in fighting against counterfeit drugs and other medicines.

Stock is developing a universal sharing network (USN) based on Blockchain and IoT for selling or renting goods securely with the transactions being tamper-proof and transparent.

Pavo, an agricultural company, implements Blockchain-based IoT solutions in their services to bring transparency and security to their farmers by using sensors to collect data and allowing the stakeholders to make an informed decision. Through smart contracts, they empower the farmers to sell with a trust system before their products are harvested.

NetObex has been implementing various IoT and Blockchain services measuring river contamination by teaming up with other companies to using drones to collect water samples, analyze their contents, and publish on Ethereum with real-time data; smart parking solutions for auto payment; and drive-thru restaurants and smart charging stations using their IoToken.

Helium, based in California, implements Blockchain-based IoT wireless internet infrastructure to secure and minimize the energy resources for running smart devices.

ArcTouch created a decentralized system, DApps, to connect to smart devices like Amazon's Alexa and Facebook's Messenger to provide services that are more secure and connected over the network.

Chronicled empowers Blockchain-based IoT solutions in the pharmaceutical and food supply chain to ensure the custody trial over its platform instilling trust and confidence among its stakeholders. Using the privacy policies of medicines is one of their highly sought-after services.

Hypr uses decentralized systems integrated into their IoT system to secure ATMs, homes, locks, and cars to secure their line of communication. Using a digital key provided to homeowners, which cannot be changed, the smart devices communicate in sync with the user's requirements.

Xage implements Blockchain-based IoT solutions with a list of DApps (decentralized applications) ranging from agriculture to supply chain in managing policies and communication among the devices, and has recently collaborated with the Smart Electric Power Alliance to focus on bringing clean energy sector to avoid cyberattacks.

Grid+ employs Ethereum to give its clients access to energy-saving IoT devices. An agent buys and manages the electricity to connect to energy-saving smart devices to pay for an optimized amount every 15 minutes, which are secured using Blockchain transactions. They are currently working on Lattice1 to determine the most efficient energy-saving point of time and integrate it into their business model.

Filament, based in Nevada, has designed products integrating Blockchain and IoT, called Block let US Benclave, for enhancing the security protocols in construction, manufacturing, and supply chain industries.

In Industry 4.0, integration of technologies becomes highly recommended and shows the most promising result in terms of convergence with cloud computing, Big Data, AI, IoT, Blockchain, Robotics, 5G, etc. There are also several types of research conducted and demonstrated in [2, 6, 7] and other papers where applications of smart home, botnet, IDoT and governance, security solutions, smart cities, privacy-preserving healthcare, and food traceability are further illustrated and elaborated as potential solutions for overcoming the limitations posed in these domains.

2.4.4 Challenges for a Blockchain-Based IoT Solution

Until now, the market for a Blockchain-based IoT has demonstrated the demand and its potential for being a breakthrough solution in many industries and domains of the economy. Many proof-of-concept or pilot runs have been in practice since 2017. Everledger, an application specific for diamond tracking and Filament for IoTs, has emerged, but if we look at real-world, end-to-end integration of Blockchain technology, the result is mostly ambiguous or restricted and not concrete or fully exploited. It becomes quite challenging to draw the line between Blockchain and IoT's potential and their limitations. This is due to the high uncertainty in its deliverance and performance.

The logical challenges of a Blockchain-based IoT described in [27] are:

1. Limited resources for IoT devices such as sensors and other nodes. Blockchain is not at all free from cyberattacks like selfish mining, DDoS, forking attacks, etc.

2. The transaction process is slow in comparison to the demand speed.

3. Concentration and consolidation of IoT devices with Blockchain's mining capabilities due to limited computing power of end-points.

4. IoT devices require the full suite of Blockchain's standard protocol of cryptographic elements.

5. Storage Capacity: for any new IoT device to enter the Blockchain platform, it is required to download all of its initial transactions, which the devices may not be able to store.

6. All the participating nodes must be a part of a P2P network with upload and exchange happening continuously.

2.4.5 Alternatives

It is also important to understand that alternatives to Blockchain and IoT do exist. To evaluate the potential of a Blockchain-based IoT solution, stakeholders and researchers should not restrict to just a few technologies. In this section, the various alternatives to these technologies are mentioned.

IoT connectivity with Wi-Fi can be viewed in parallel to cellular connectivity (satellite connection), which provides one of the farthest ranges of the network (around 10-15 miles using a broadcast tower) unlike Wi-Fi. The low-power wide area network (LPWAN) consumes very little power and offers a new look over the IoT network (LTE-M, NB-IoT, LoRa). Zigbee is used in Amazon Echo. Z-Wave uses a 908 MHz bandwidth while reducing the interference. Network connectivity is an important factor that determines the global standard for IoT-based applications.

1. IOTA: IOTA [28] open-source distributed ledger and cryptocurrency designed for the Internet of things, which is based on DAG technology, is gaining popularity among its competitors. Directed acyclic graphs (DAG) record transactions on a DLT, allowing cryptocurrencies to be independent of Blockchain. DAG was proposed by Sergio Demian Lerner in 2015 as an alternative to Blockchain. It eliminates the need for miners and still works in a distributed, peer-to-peer, decentralized manner by validating transactions with new transactions. Despite its hype, DAG is still in its infancy and requires a lot more work to be conducted to challenge Blockchain services. IOTA is basically a protocol with a Tangle DTL that works without the need for mining. It comes under the infrastructure for an IoT-based ecosystem that analyzes data from its devices. Tangle ledger

provides IOTA the support similar to what a Blockchain does but without the need for transaction fees. It is already being supported by many enterprises.

2. Hash Graph Technology: Hash graph technology is another alternative to a distributed ledger, invented by Leemon Baird. It eliminates the constraints of low speed and the need for validation of transactions by Blockchain technology. Instead, the transactions are timed using DAG.

3. HoloChain Technology: Holo chain technology is another recent advanced technology that employs cryptographic elements to develop secured applications in a decentralized P2P network. It allows multiple storage of data within subchains and eliminates the need for a consensus algorithm, unlike Blockchain.

4. R3 Corda: Corda is another digital ledger that records financial transactions using a P2P model. It queries multiple nodes in the chain. It works on a set of designated peers that can be trusted, thus providing security to the transactions. Corda is a simplified version of DAG in terms of its smart contract design and functioning.

5. Ambrosius: Ambrosius is a public permissioned Blockchain that works in a decentralized network. It is an optimized version of Blockchain for interconnectivity of an IoT system and management software mainly for tracking and integrating Blockchain into a business model.

6. Atonomi: Atonomi is an open platform supporting trusted third parties for IoT stakeholders an embedded solution to secure devices with Blockchain's immutable registration of identity and traceability features. The IoT devices can have integrity over their communication and provide a different approach.

7. Chain of Things (COT): COT is a consortium of stakeholders who are interested in Blockchain-based IoT solutions to improve upon its challenges such as security, privacy, limited resources, scalability, etc.

8. Modium.io: Modum.io is a Blockchain-based IoT solution for tracking and monitoring supply tangible goods using IoT sensors to check on the goods while maintaining a smart contract based on Blockchain to exploit its immutability, transparency, and security features.

9. Xage: Xage is the first cybersecurity platform based on Blockchain for IoT enterprises that can manage billions of devices at a time to secure and self-diagnose any attacks. It is being used mostly in the supply chain and energy industries.

2.5 FUTURE TRENDS AND RESEARCH AREA

Research shows that future direction for improvement can be observed in line with real-world practical implementation at a global standard to solve the privacy and security features, beyond the known practices for metadata [14, 28]. These include the challenges faced in adopting the large storage capacity for Blockchain's need of duplicated data in the participant nodes for IoT's limited resource capacity. Conclusion

The chapter can be concluded with an understanding of the various shares of the market held by Blockchain- based IoT solutions. Getting an insight into how Blockchain works is essential, as it represents the engine that will be powering most of the future applications or at least what is expected by the economy's current needs, and the future needs can be outlined on the basis of this study. The technical aspects of each of the emerging technologies shed some perspective on the recent and current work being conducted. From its challenges to benefits, the platforms described can be established as a value-added solution that has the potential to scale and deliver the applications' various needs. Future research also needs to focus on where and what to deploy and develop to achieve a global standard that can be accepted to bring its opportunities into fruition.

REFERENCES

[1] J. Xu, K. Xue, S. Li, H. Tian, J. Hong, P. Hong, and N. Yu, "Healthchain: A blockchain-based privacy preserving scheme for large-scale health data," *IEEE Internet of Things Journal*, vol. 6, pp. 8770–8781, 2019.

[2] J. Lin, Z. Shen, A. Zhang, and Y. Chai, "Blockchain and IoT based food-traceability for smart agriculture," in *Proceedings of the 3rd International Conference on Crowd Science and Engineering-ICCSE'18*, 2018.

[3] S. Hakak, W. Z. Khan, G. A. Gilkar, M. Imran, and N. Guizani, "Securing smartcities through blockchain technology:Architecture, requirements, and challenges," in *IEEE Network*, vol. 34, no. 1, pp. 8–14, 2020.

[4] S. Dorri, S. Kanhere, R. Jurdak, and P. Gauravaram, "Blockchain for IoT security and privacy:The case study of a smart home," in *2017 IEEE International Conference on Pervasive Computing and Communications Workshops (PerCom Workshops)*, 2017, pp. 618–623.

[5] M. A. Khan, and K. Salah, "IoT security: Review, blockchain solutions, and open challenges," *Future Generation Computer Systems*, vol. 82, pp. 395–411, 2018.

[6] O. Novo, "Blockchain meets IoT: An architecture for scalable access management in IoT," *IEEE Internet of Things Journal*, vol. 5, no. 2, pp. 1184–1195, 2018.

[7] H. F. Atlam, and G. B. Wills, "Technical aspects of blockchain and IoT," in *Advances in Computers*, Elsevier, pp. 1–39, 2018.

[8] J. Gubbi, R. Buyya, S. Marusic, and M. Palaniswami, "Internet of things(IoT): A vision, architectural elements, and future directions," *Future Generation Computer Systems*, vol. 27, no. 7, pp. 1645–1660, 2013.

[9] H. Liu, D. Han, and D. Li, "Fabriciot: A blockchain-based access control," *IEEE Access*, vol. 8, pp. 18207–18218, 2020.

[10] A. M. Saghiri, M. Vahdati, K. Gholizadeh, M. R. Meybodi, M. Dehghan, and H. Rashidi, "A framework for cognitive internet of things based on blockchain," in *2018 4th International Conference on Web Research (ICWR)*, 2018, pp. 138–143.

[11] B. Rodrigues, E. Scheid, C. Killer, M. Franco, and B. Stiller, "Blockchain signaling system (bloss): Cooperative signaling of distributed denial-of-service attacks," *Journal of Network and Systems Management*, vol. 28, no. 4, pp. 953–989, 2020.

[12] A. Kosba, A. Miller, E. Shi, Z. Wen, and C. Papamanthou, "Hawk: The blockchain model of cryptography and privacy- preserving smart contracts," in *IEEE Symposium on Security and Privacy (SP)*, 2016, pp. 839–858.

[13] Y.-C. Liang, *Blockchain for Dynamic Spectrum Management*. Singapore: Springer Singapore, 2020, pp. 121–146.

[14] Z. Zheng, S. Xie, H. Dai, X. Chen, and H. Wang, "An overview of blockchain technology:Architecture, consensus, and future trends," in *2017 IEEE International Congress on Big Data (Big Data Congress)*, 2017, pp. 557–564.

[15] S. Hakak, W. Z. Khan, G. A. Gilkar, M. Imran, and N. Guizani, "Securing smart cities through blockchain technology: Architecture, requirements, and challenges," *IEEE Network*, vol. 34, no. 1, pp. 8–14, 2020.

[16] M. A. Khan and K. Salah, "IoT security: Review, blockchain solutions, and open challenges," *Future Generation Computer Systems*, vol. 82, pp. 395–411, 2018.

[17] N. Szabo, "Formalizing and securing relationships on public networks," *First Monday*, vol. 2, no. 9, 1997.

[18] T. Ahram, A. Sargolzaei, S. Sargolzaei, J. Daniels, and B. Amaba, "Blockchain technology innovations," in *2017 IEEE Technology Engineering Management Conference (TEMSCON)*, 2017, pp. 137–141.

[19] I. Lee and K. Lee, "The internet of things (IoT): Applications, investments, and challenges for enterprises," *Business Horizons*, vol. 58, no. 4, pp. 431–440, 2015.

[20] S. Huh, S. Cho, and S. Kim, "Managing IoT devices using blockchain platform," in *2017 19th Inter- national Conference on Advanced Communication Technology (ICACT)*, 2017, pp. 464–467.

[21] M. J. M. Chowdhury, A. Colman, M. A. Kabir, J. Han, and P. Sarda, "Blockchain versus database: A critical analysis," in *2018 17th IEEE International Conference On Trust, Security And Privacy In Computing And Communications/12th IEEE International Conference On Big Data Science And Engineering (TrustCom/ BigDataSE)*, 2018, pp. 1348–1353.

[22] M. Humayun, N. Jhanjhi, B. Hamid, and G. Ahmed, "Emerging smart logistics and transportation using IoT and blockchain," *IEEE Internet of Things Magazine*, vol. 3, no. 2, pp. 58–62, 2020.

[23] J. Gubbi, R. Buyya, S. Marusic, and M. Palaniswami, "Internet of things (IoT): A vision, architectural elements, and future directions," *Future Generation Computer Systems*, vol. 29, no. 7, pp. 1645–1660, 2013, including Special sections: Cyber-enabled Distributed Computing for Ubiquitous Cloud and Network Services & Cloud Computing and Scientific Applications — Big Data, Scalable Analytics, and Beyond.

[24] N. Živi, E. Kadušić, and K. Kadušić, "Directed acyclic graph as tangle: An IoT alternative to blockchains," in *2019 27th Telecommunications Forum (TELFOR)*, 2019, pp. 1–3.

[25] A. Bellini, E. Bellini, M. Gherardelli, and F. Pirri, "Enhancing IoT data dependability through a blockchain mirror model," *Future Internet*, vol. 11, no. 5, p. 117, 2019.

[26] R. Bruno, S. Eder, K. Christian, F. Muriel, and S. Burkhard, "Blockchain signaling system (bloss): Cooperative signaling of distributed denial-of-service attacks," *Journal of Network and Systems Management (JNMS)*, vol. 28, pp. 953–989, 2020.

[27] F. Buccafurri, G. Lax, S. Nicolazzo, and A. Nocera, "Overcoming limits of blockchain for IoT applications," in *Proceedings of the 12th International Conference on Availability, Reliability and Security*. Association for Computing Machinery, 2017.

[28] A. Panarello, N. Tapas, G. Merlino, F. Longo, and A. Puliafito, "Blockchain and IoT integration: A systematic survey," *Sensors*, vol. 18, no. 8, p. 2575, 2018.

CHAPTER 3

A Review on Blockchain-Based Device Authentication Schemes for IoT

G. Megala
Vellore Institute of Technology, Vellore, Tamil Nadu, India

Prabu Sevugan
Vellore Institute of Technology, Vellore, Tamil Nadu, India
Department of Banking Technology, Pondicherry University, Pondicherry, India

P. Swarnalatha
Vellore Institute of Technology, Vellore, Tamil Nadu, India

CONTENTS

DOI: 10.1201/9781003188247-3

3.1 INTRODUCTION

A vast network that consists of interconnected devices like smart phones, computers, sensors, and actuators is known as the Internet of things (IoT), which has evolved exponentially in recent years. It has the potential to allow modern devices to determine themselves and communicate with one another. These smart devices would be able to collect, evaluate, and even make decisions without the need for human interaction or social contact. Despite the abundance of IoT devices and their deployment in sensitive applications, IoT-related security risks have become much more concerning. There are many security terminologies and definitions that are widely utilized, including different security aspects:

- Message: A set of bytes exchanged between two parties.

- A shared secret is often a value that is shared by two entities. The code may be as easy as a password or an encryption key that all parties are aware of.

- Confidentiality or secrecy: A method of prohibiting an unauthorized investigator from assessing the contents of the message.

- Integrity: A criterion for determining whether or not a message has been tampered during the transmission and storage.

- Authentication: A method of implying that a message can be traced back to the sender, allowing the receiver to check that the message was sent only by the sender.

- Authorization: Evidence that the consumer is authorized to perform a task.

- Nonrepudiation: A method of preventing a message's owner from arguing that the message was not sent.

- Resistant replay: A method of blocking a message from being reused by an attacker.

- The performance standard security measures of an IoT network are largely determined by the software implementations it supports; there is a need for secrecy, privacy, honesty, and authenticity, which is directly proportional to the application or software's security compliance requirements.

3.1.1 Authentication

When data moves through an unsecured variety of social media platforms, IoT authentication is indeed a paradigm for establishing trust in the identification of IoT computing devices in order to maintain security and manage access. IoT systems have limited resources and are unable to secure or protect themselves, making them rather easily compromised. As a result, adequate authentication and authorization schemes must be implemented to ensure optimal security of IoT devices, communications, and information. The authentication mechanism must be trustworthy [1], flexible, and resistant to proven cyberthreats [2]. Unauthorized third parties try to access the stored information such as confidential data, media files, and interactions available on the server. Authentication approach detects such entities and denies them access. Figure 3.1 illustrates the different types of authentication.

There are primarily three types of user authentication approaches: (1) knowledge-centered authentication describes what the client knows and wants (examples: password, security questions, OTP); (2) possession-based authentication specifies what the client possesses or holds (examples: One time pad through mail or text, software tokens, access certificates); and (3) inherence- or biometric-based authentication defines who the user is (examples: fingerprints, iris or retinal images, biometric, facial, voice recognitions). Other factors include time factor authentication (attempted login, permitted and rejected access), location factor authentication (which identifies the accessing device using a global positioning system [GPS] to confirm identity), and artifacts-based authentication (which involves smart cards, digital signatures, and certificates).

To secure communications between IoT devices, a range of methodologies can be used to establish strong authentication, such as:

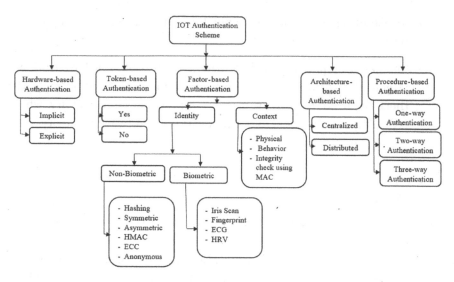

FIGURE 3.1   Authentication schemes.

- Single-mode authentication – either one of the communicating devices will be authenticated.

- Double way or mutual authentication – both devices involved in communications will be authenticated.

- Three-mode authentication – the central or third-party authority authenticates other two users or devices.

- Authentication in a distributed manner.

- Authentication in a centralized manner.

Often these IoT authentication techniques involve a trustworthy third party to authenticate the users using the open authorization (OAuth) protocol, which is centralized in design and implementation. This method has many disadvantages, including a significant expense, a single point of failure, intrusion, and potential privacy breach.

IoT systems are often compromised externally, with a hacker attempting to gain access to the device via an internet service. Any such efforts to interact with an IoT computer would be ignored if it is only enabled to interact with an authenticated server. As per the 2018 Symantec threat survey, the incidence of IoT attacks increased from 6,000 in 2016 to 50,000 in 2017. As a result, while IoT devices are deployed within enterprise networks, protection must be prioritized. To tackle this limitation, cryptography approaches that are both effective and efficient should be used to standardize safe communication among devices. Choosing the best IoT authentication model for the project, on the other hand, is a difficult task. Until determining which implementation model is more appropriate for IoT authentication, users must consider a variety of factors, including energy resources, infrastructure capability, affect income, security experience, security measures, and functionality.

3.1.2 Device Identity Management

Device identity management seeks to increase and retain confidence in a device's identity as it communicates with other computers, apps, clouds, and access points. Industrial control systems [3], connected consumer healthcare, automotive ECUs (Engine Control Units), surveillance cameras, home security systems, smart phones, and smart appliances are examples of IoT devices that may require authentication and authorization. While linking to a server, each IoT device requires a unique identifier to prohibit fraudsters from acquiring access to the device. This is done by associating an identity with a secret key generation algorithm that is unique to each IoT unit. The device chooses the secret key at random using elliptic curve cryptography (ECC) [4] so as to provide high-level security, and the corresponding device uses the generator G point of elliptic curve, which is repeated a number of times, thus generating a distinct public key. Provided just a public key, finding the hidden key pair within polynomial time is nearly impossible, as it is based on the elliptic-curve discrete logarithm problematic (ECDLP) [5] computation, which is non-polynomial. As the hash functions are irreversible, they can be used to generate address, and so it is impossible to get the public key based on the address. The registration ID is generated by the trusted platform module (TPM) for TPM-based implementations. An internationally trusted Certificate Authority (CA)produces the registration ID for X.509 certificates. Device identity management strategies are in charge of identifying and managing the privileges or credentials used by devices during their entire lifespan.

The X.509 protocol is focused on the certificate chain of trust, model and provides the most reliable digital identity authentication method. Using X.509 certificates as an authentication mechanism is a great way to increase efficiency and make equipment distribution easier. Moreover, due to the technical challenges involved, X.509 certificate lifecycle management can be difficult and costly, contributing to the total solution expense.

As a result, many customers depend on third-party providers for certificates and lifecycle management.

The hardware security module (HSM) is the safest type of secret storage, since it is used for stable, hardware-based system confidential storage. The HSM can store both the X.509 certificate and the SAS tokens. HSMs are compatible with the provisioning service's two attestation mechanisms. Device keys and credentials can also be stored in the system memory, although this is a less reliable method of storage than that of an HSM.

3.1.3 Device-Based Authentication

Device-based authentication and authorization would most likely be used by devices whose connections are not dependent on the operator. A car is a good example: whether or not the car has user-specific communication—such as executing user applications or having user-specific configurations—the system (i.e., the car) would most probably get to be linked to share vehicle specifics [6]. User certificates stored in the car are used for device-based authentication and authorization. During the TLS handshake, authentication generally occurs. The car (device) would start sending its certificate, which will then be signed by a configurable signing authority, as well as details authenticated by its secret private key during the handshake. The device is authenticated after completion of the handshake, and the user ID is obtained from the certificate. The user ID for a car (device) may be the vehicle unique identification number (VIN). After the car has been authenticated and the user ID has been determined, the correct authorization sections must be determined. A system can enter an authorization community simply by connecting. Since the unit in this case is a vehicle, all cars that attach will be assigned to the authorization category car. Additional classes, much like the client ID, can be generated from the certificate. The VIN of the vehicle [7], for example, can be used to extract a number of classes. Unmanned aerial vehicles in a smart city are linked together via peer-to-peer networks and a shared ledger [8]. A public Blockchain can be used to allow all drones to register to fly on approved routes after authenticating on the Blockchain. Via their disseminated identity, all linked devices are allowed to migrate within certain regions. Those devices that are authenticated once no longer need to be authenticated again when they move or enter a different zone. There is a chance of occurrence of authentication delay and authentication attacks during the communication process. Major concern is about geographical area range changeover and device speed, which affect the performance of devices equipped with an IoT architecture.

3.1.4 Need for Device Authentication

A perfect robust device authentication is necessary to guarantee that all devices connected to a distributed IoT network are trustworthy. As a result, every device in the network has its distinct identity, which is verified when those devices are connected to the central server or access point of the distributed systems. A network administrator can monitor each computer with its unique identity during its development process, interact safely, and avoid performing harmful processes. Network admins can simply terminate a device's permissions if it demonstrates unusual conduct. A powerful authentication is required to safeguard devices and sensors associated with the IoT network being accessed by users who are not authorized to do so. IoT devices rely heavily on Blockchain technology [9] for management, control, and, most importantly, security. Blockchain has the potential to be a key enabler in delivering viable security solutions to increasingly complex IoT security issues.

3.1.5 Authentication Protocols

To secure IoT devices, IoT authentication methods are required, and there are several options for accomplishing this goal. In the name of protection, some organizations can delegate dedicated IoT networks, sacrificing useful features. It is possible to incorporate it in a safe way, but it needs to be ensured that devices are designed with security. Nevertheless, since IoT devices use a broad range of protocols and standards, authentication mechanisms must account for a wide range of variables.

As a result, expertise with those kinds of modifications is needed, as is expert understanding of the IoT devices to ensure that each device is capable of safe authentication [10, 11]. Some might need some manual updating (due to a lack of OTA operability), whereas others might have restricted configurations that cannot be modified. Figure 3.2 represents the different protocols used at different layers of the network. To create the desired degree of confidence, deprived of overloading the model, it is critical to examine the elements necessary for identity verification of knowledge possessions. Although standard protocols that perform authentication, such as public key infrastructure (PKI) and open authorization (OAuth), are increasingly being amended/updated to serve the necessary scope and security, the resource-controlled existence of several IoT endpoints

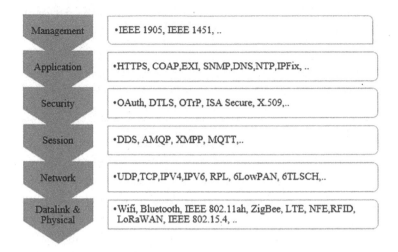

FIGURE 3.2   IoT protocols.

significantly limits its capacity to support popular authentication methods, leading to the market acceptance of standardized authentication mechanisms that do not comply with trust criteria and provide very restricted or no interconnectivity. Fortunately, this is increasingly evolving as the industry shifts from treating protection as an academic exercise to incorporating it into the device design phase.

Mainstream integrated devices nowadays use a software-based solution for device authentication. Utilization of message authentication code (MAC) allows alteration of safe secret key to authenticate devices in restricted IoT procedures; PKI implementations and OAuth conventions are two common examples. OAuth is one of the most common and widely used authentication mechanisms, particularly for IoT devices. OAuth is an open-standard protocol for token-based authentication and authorization that allows users to access sensitive information via a trustworthy centralized server [12, 13]. Figure 3.3 depicts the main actors involved in the OAuth protocol, as well as the main steps those actors take.

The following four actors are involved in the protocol:

- The resource owner is the person who owns the protected resources and has the authority to grant access to them.

- The authorization server (AS) is responsible for distributing access tokens to approved clients.

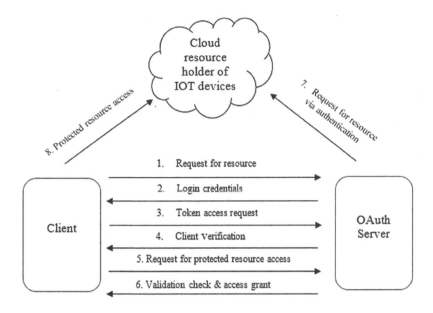

FIGURE 3.3 Overview of OAuth authentication system.

- The resource server that hosts the secured resources and accepts access based on the AS's issued tokens.

- The client who makes the request for entry through an application software.

First, the client uses the authorization server to seek permission from the resource owner. Second, the authorization server submits the login credentials to the client. In the third step, the client authenticates oneself to the authorization server using the login credentials obtained from the AS to request an access token from the AS. The AS verifies the user and provides the client with an access token in the fourth step. In the fifth step, the client uses the obtained access token to request access to secure resources from the AS. The AS authenticates the access token in the sixth step, and if the validation is correct, the client is permitted to access the protected services.

While software-based IoT authentication offers significant maintenance costs benefits, the security provided through this strategies is severely limited by means of privacy of the framework of the device developers. IoT authentication and authorization schemes are being implemented in the growing generation of IoT system, providing a defense mechanism against the breaches. The functionality of IoT systems is known to be ineffective against popular application attacks and hardware quenching. Machine hardware–centered security approaches are quickly emerging to meet the industrial standards in protecting PCs, notebooks, and smartphones based on trust execution. These methods have become more important in terms of securing IoT computers. In terms of chip security, mechanisms such as key pair generation, integrity protection, and acknowledgment are performed in an integrated hardware environment, providing effective protection from external theft. When combined with application PKI, the trusted platform module (TPM) is a perfect illustration of root of trust deployment and provides the highest trust authentication for IoT devices. TPM confidence assessments provide safe command in booting connected devices and authentication of each and every device connected to the network, in addition to preservation of digital credentials.

The identity validation parameters differ among each and every device, involving device sustainability phases, and the potential consequences compromise through unauthorized access. This drives selecting an acceptable authentication mechanism for those sets of interconnected devices. The confidence level provided by available system authentication mechanisms should be correlated with the trust parameters of IoT devices by IoT security design professionals. TPM and other hardware-based authentication mechanisms are especially suitable for IoT use-cases where ensuring higher levels of confidence in the user identity is critical.

3.2 PRELIMINARIES

3.2.1 Public Key Infrastructure Cryptography

Public key infrastructure is the asymmetric cryptographic approach mostly used for encryption purpose for the communication that takes place over the internet. The public cryptographic keys at its core, as the name suggests, are a crucial concept involved. These keys are used not only to encrypt data but also to verify the identity of the communicating entities. The most distinguishing characteristic of public key infrastructure is that it attains the inherent security provision using public key, which is disseminated to the public and a private key known only to the user. Areliable infrastructure is needed for secure key exchange management and maintenance. For public key cryptography, there are two key management criteria.

Private Key Confidentiality – Private keys essentially endure undisclosed among whole entities apart from the data owner and allow utilization of the key during its lifecycle.

Ensure Security of Shared Keys –Public keys are known to everyone, and any sender can use those receivers' public key to encrypt the data to be communicated securely without sharing the private keys. This encryption using receiver's public key is based on discrete logarithmic problem, as it is difficult to find the private key of the sender. Moreover, it also is not guaranteed that the accompanying public keys will be accurate. As a result, public key management [14] must place a greater emphasis on ensuring that public keys are used for their intended purpose.

PKI ensures the security of shared keys. It allows for the identification and distribution of public keys. The following components make up the PKI composition.

- The term "digital certificate" refers to a public key certificate.

- Tokens for private keys.

- Certification Authority (CA).

- Registration Authority (RA) – The authority in charge of registration.

- System for managing certificates.

(i) Digital Certificate

A certificate can be compared to identification card of a person, such as a driver's license, voter ID, or a permit, in terms of proving one's identity. Digital certificates are given not only to individuals but also to computers, software packages, and everything else that requires proof of its distinctiveness in a massive environment.X.509 is a digital certificate based on the International Telecommunication Union (ITU) X.509standard. It specifies a regular format of a certificate to be issued for public key infrastructure, authentication, and authorization of certificates.

The certification authority (CA) or expert holds the global public keys pertaining to the user account in digital certificates, including further related information of the client and its usability. The whole information is digitally signed by CA, and the certificate contains a digital signature. Anyone who requires confirmation about a client's public key and related information performs signature validation using the public key of CA. Validation ensures that the certificate's public key belongs to the individual whose information is included in the certificate.

(ii) Tokens

If a client's public key is stockpiled within digital certificates, hidden private key accompanying it may be saved in the client's device. Storing keys is not common practice. An intruder who gains access to the machine can easy open and use the private key. As a result, the private key is held on an encrypted removable storage token that requires a password to access. When it comes to storing keys, various vendors use a variety of formats, some of which are licensed.

(iii) Certificate Authority (CA)

The CA disputes a certificate for each customer as well as helps other clients validate it. The CA is responsible for properly identifying the client requesting a certificate, as well as ensuring that the details provided within the certificate are appropriate and signing is done. The following are the main functions of a CA:

Creating a Pair of Keys – CA can create a key pair on its own and/or in collaboration with the user.

Creating and Delivering Digital Certificates – CA can be compared to a passport agency in terms of PKI. After the client presents the credentials to validate its distinctiveness, CA then issues a certificate, which is then signed by the CA to keep it from being altered.

Certificate Dissemination – When the client or user is being located, CA publishes the client's certificate. This can be accomplished either by disseminating the certificate in publicly available certificate directory or by providing it only to a particular client.

Certificate Authentication – CA allows its public key accessible to aid in the sign verification in the authentication process of clients' certificate.

Cancellation of Certificate – CA can remove a certificate issued for a variety of reasons, including the user's compromise of the private key or a lack of confidence in the client. After removal, the CA keeps a registry of all revoked certificates that are still valid.

(iv) Registration Authority (RA)

CA may enlist the help of a third-party RA to conduct the required identity verification authorizations of individual and/or business clients' requests for a certificate. Although the RA appears as certifying authority to the client, it does not really sign certificates.

(v) System for Managing Certificates (CMS)

Certificates are released, renewed, suspended permanently/temporarily, or withdrawn using the CMS. CMS often does not revoke certificates, as they might be required to prove its eminence at some place in the future (e.g., legal issues). To be able to monitor their responsibilities, a CA and its related RA run certificate management systems.

With large networks and global communications criteria, having a single trustworthy CA in which each requesting user receives a certificate is practically impossible. Moreover, if only one CA is accessible, it can cause problems if that CA is compromised. The method of guaranteeing the certificate chain is legitimate, appropriately signed, and reliable is known as certificate chain verification. The protocol that follows verifies the certificate chain, starting with a certificate that is provided for authentication.

Clients' certificate is supplied to the chain of certificates as when the clients are validated and is forwarded to the root of the authority. The provider public key, which can be found in its certificate, is used by the verifier to verify the certificate. It is found next to the client's certificate in the certificate chain. If the verifier trusts that the providers' certificate is signed by an appropriate authority, verification is complete and the process ends here. Alternatively, the providers' certificate is validated in the same way as the client's certificate was verified in the preceding measures. This procedure is further continued until finding a trustworthy CA in the middle, or until root CA is found.

3.2.2 Multi-Factor Authentication (MFA)

MFA is a substantiation scheme that permits a user to take several verification parameters to gain access control on the resources such as a software, an online account, or a virtual private network. A successful identity and access management (IAM) approach essentially affords MFA. MFA needs several extra authentication measures, including login details, which reduces the risk of a potential security breach. MFA operates by requesting extra information for verification factors. One-time passwords (OTP) are among the most important MFA variables that users experience. OTPs are four- to eight-digit codes that you can receive via text, e-mail, or through any smartphone application. When using OTPs, a unique code is provided on a regular basis or whenever an authentication request is made. The code is created using a seed value assigned to the client when they first register, as well as another factor, such as an incremented counter or indeed a time value.

3.2.2.1 Multi-Factor Authentication Methods

Authentication approaches are becoming more complicated as MFA incorporates AI technology and intelligent automation. The majority of MFA authentication methods depend on the first three forms of additional data. Those are knowledge-founded authentication, biometric-centered authentication, possession- or inherence-focused authentication, time-based authentication, and location-based authentication. OTPs can be both knowledge-based as well as possession-based authentication of users and devices in the network. Based on location, MFA usually examines a device's IP address and, if accessible, its geographical location. This detail can be used to block a user's access when the location data does not fit what has been listed on a block list, or it can be used in conjunction with other forms of authentication, including a password or an OTP, to validate the user's identity.

Adaptive authentication, also known as risk-based authentication, is a subset of MFA. When authenticating, adaptive authentication considers additional variables such as perspective and behavior, and these values are often used to assign a degree of risk to the login session. The level of risk is assessed to decide whether a user is able to log in or whether an additional authentication factor is needed. With adaptive authentication in effect, a user logging in past midnight from a cafe, which is not something they usually do, could be asked to enter a code texted to their smartphone in addition to their login credentials.

3.2.3 Integrity Verification

Owing to a centralized approach, conventional data integrity strategies including symmetric key cryptographic methods [15] and public key infrastructure (PKI) cryptosystem struggle with hardware failures and transmission delay in a congested network. Data can be tampered with at intermediate nodes using symmetric key solutions. The sender generates a MAC code for the transmitted message using a mutual secret key and attaches the code with the transferred content in the symmetric-key based strategies [1, 5]. Using identical shared secret key, the verifier will check the message's authenticity on the received content. Figure 3.4 represents the data integrity verification mechanism. The key management, on the other hand, is incredibly challenging. The flexibility of symmetric key–driven applications is very low as the size of the network grows, because each communicating parties must exchange a unique secret shared key.

Furthermore, only the community of communicating nodes that have access to the secret shared key are able to verify the information validity, so this method is used in several hop connections, which is being restricted by a common type of network. The security of symmetric key–based approaches is adversely affected even if an individual unit in a group is compromised. In message authentication, there are two forms of privacy: conditional and unconditional. The exact communication sender may be detected by a trusted third party (TTP) or group administrators in conditional privacy. In unconditional message authentication, no one can recognize the truc message sender, which represents absolute privacy.

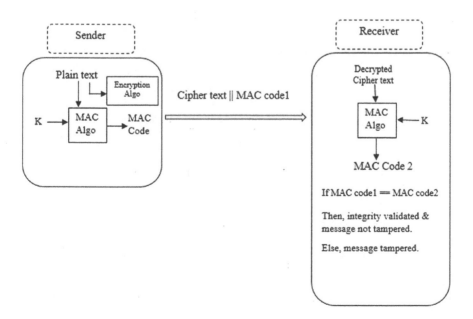

FIGURE 3.4   Data integrity verification scheme.

Asymmetric key algorithms with private and public key pairs can also be used to enforce a data integrity approach. The actual data is first encrypted by a source node using a private key. The receiver node will then decrypt the data using the source node's public key. The privacy of obtained data can be assured by relating decrypted data to actual data. The asymmetric PKI-centered technique is used to authenticate the public key to protect it from being manipulated [15]. In register authority (RA), a source node must register its distinctive identity details as well as public key information, where those details are being used to create certificates by a certificate authority. CA will create a certificate for the source node's identity using its public key after auditing its identity in RA. CA produces a pair of asymmetric keys based on public key cryptography to perform encryption on the certificate using the private key. A digital signature is defined as an encrypted certificate. Both nodes in the network have access to the corresponding public key. CA issues the certificate to RA. RA then sends the certificate along with the original and encrypted information to the audited source node through the intermediate nodes. The certificate is transmitted to the intermediate node with the real data and the encrypted data. The intermediate node then requests an equivalent public key from CA. Once the public key is received by intermediate nodes, it then accomplishes subsequent steps to verify the data integrity. (1) The digital signature would be decrypted using the public key to detect potential certificate tampering. The intermediate node will acquire the source node's public key along with the certificate. (2) To ensure confidentiality of original information, this node relates it to the decrypted information, which is being decrypted using the source nodes' public key.

3.3 BLOCKCHAIN TECHNOLOGY

In untrusted network settings, Blockchain can be defined as a disseminated public ledger scheme that protects the records of every transaction [16]. A Blockchain is made up of several blocks that are connected together forming a chain structure and are secured by using cryptographic techniques, and new blocks can be added to the existing chain at any time. The Blockchain consists of several blocks/nodes, but they are not expected to believe one another; if adequate blocks/nodes are truthful, the Blockchain can be assured to be secure [17]. Blockchain protects the network's trust in a decentralized flat structured manner, as opposed to a categorized arrangement in integrity checking. Each block consists of head and body sections. The head section consists of hash value, previous header hash value denoted by pre-hash, time stamp, and nonce value. The body section consists of a list of transactions denoted by Tr, which are being represented in tree structure connected to a merkle root. Most nodes may verify the validity of these transaction data. The structure of Blockchain is shown in Figure 3.5.

Blockchain is an immutable record that is built in a decentralized manner without the need for central authorization. Each node that involves calculation is represented as a member of the Blockchain. As these nodes validate the transactions, these are miner nodes in a process known as mining. These miners use a smart contract for validating transactions and generate blocks having an effective collection of transactions. These miner nodes make decisions using a consensus algorithms [18] such as work-proof, stake-proof, proof of activity, burn, elapsed time, etc. Different forms of trustworthy chains have been added to Blockchain since Bitcoin's introduction, such as public permissioned and permissionless blockchain

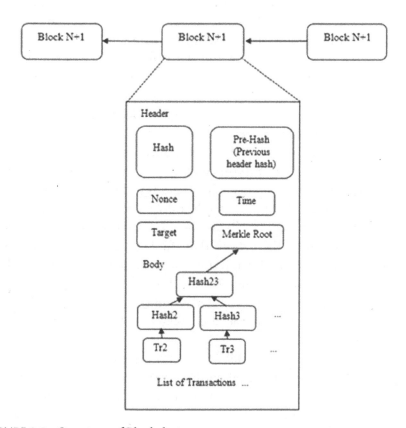

FIGURE 3.5   Structure of Blockchain.

[19, 20]. Such Blockchains are built on the concept of giving each participant unique permissions to perform particular functions. Anyone without a specified identity will participate in mining on a public Blockchain. Local cryptocurrencies are typically used in public Blockchain, which use economic rewards and consensus mechanisms such as work proof and deputized stake proof. A secret permissioned Blockchain is one that is completely private or restricted to a small number of allowed nodes. Permissioned Blockchain uses a collection of identified unique participants to execute the Blockchain and provides a way to ensure cooperation between a group of people with shared interests but not complete trust on one another. Permissioned Blockchain is restricted to a small number of registered users, allowing them to form a network. Several administrations may enter the network by getting their individual representatives.

Every node in the network is given two keys, a public and a private one, which are employed in encryption and decryption processes for secure message transmission. After completing a transaction, a node publishes the signed document to distinct neighbors. Every transaction is signed individually, which allows it to be authenticated using only particular signatures based on the private key, and so the integrity is ensured with the information not allowed to be decrypted if there occurs an error in the data transmission process. When the node counterparts broadcast those obtained transaction that are being signed, its legitimateness is verified before transferring with further delegates, helping the transaction propagate across the network. Special nodes termed *miners* order and bundle the transactions distributed in this way that are considered legitimate by the network into a time-stamped chain. An agreement procedure is used to choose the miners along with that the information will be encompassed within blocks. The miner's blocks are then released back into the network. The Blockchain nodes then check that the broadcasting node includes legitimate transactions and that it uses the corresponding hash to link to the previous block in the chain. The block is destroyed if these requirements are not satisfied. Moreover, if all conditions are met, the blocks of nodes are being appended to the chain when the transactions in a block are updated.

Blocks are permanent thanks to the encryption algorithm, which means they cannot be tampered with once they have been connected to the chain. The Blockchain framework relies on a consensus mechanism to keep the network running, which is managed by all nodes, which can be any computer. The consensus approach ensures that even if one node fails, the remaining nodes will continue to function normally, overcoming the drawbacks of the conventional centralized mode, which is vulnerable to malicious attacks and tampering. Blockchain is a form of a credit system that differs from the conventional distributed network's trust model of authorized intermediary acceptance. It is a no-trust scheme in a sense. Perhaps the most commonly used Blockchain that is known to be fully decentralized is the public Blockchain. A consortium Blockchain is one in which the consensus process is managed by a group of preselected nodes. This form of Blockchain is referred to as temporarily decentralize. Private Blockchain applies to the need for Blockchain only for record-keeping purposes and is not open to the public.

The advantages of Blockchain are as follows:

- Blockchain can be used to store data from IoT devices.

- Blockchain's distributed nature allows for safe data storage.

- The hash key is used to encrypt data, which is then checked by miners.

- Data loss and spoofing attacks are avoided.

- Unauthorized access can be prevented using Blockchain technology.

- For resource-constrained computers, a Blockchain-based proxy-based architecture is used.

- It has high operational efficiency, increases cyber resiliency, transparency, and global accessibility.

- Provides decentralization, traceability, accountability, and security and reduces human error.

3.3.1 Blockchain-Based Authentication

An important function involved in Blockchain-based authentication is using hash operations while signing every transaction present in the Blockchain. As a result, several applications of IoT deploying hash functions must remain stable; meanwhile it must be fast and use as little energy as possible without producing any conflicts [21]. Cryptocurrencies such as Bitcoin and name coin use the most popular SHA-256d hash function; Emercoin also uses SHA-256 hash function, whereas Litecoin and Dodgecoin uses Scrypt for authentication purpose. SHA-256's output has been tested in a variety of IoT products, including wearables. Researchers who looked at the fingerprint and resource demands of SHA-256 in ASICs concluded that the advanced encryption standard (AES)is more effective for low-power protected communications. Other authors recommended using ciphers because of their power limitations, and several applications require analytical assessments of Blockchain-based IoT.

Any business with numerous actors requires a standardized data format that can be read, updated, and used to make decisions. If a transaction is made in the Blockchain, a current timeline is registered, and no more changes are permitted after that point. Conventional time-stamping relies on a trusted centralized server signing and time-stamping transactions using private secret key. However, a compromised server has a chance of signing previous transactions. While time-stamping can indeed be decentralized, it is vulnerable to Sybil attacks, which Blockchains such as Bitcoin avoid by associating blocks in addition of employing a work proof authentication scheme.

Transactions must be authenticated and time-stamped in order to monitor changes on its Blockchain. Since this last step must be completed in a coordinated manner, time-stamping servers are frequently used. There are a variety of time-stamping methods that can be used. Conventional schemes depend on the server's truthfulness, which uses its own private key to sign and time-stamp transactions. Despite this, nothing could stop the server from signing previous transactions. As a result, a number of writers have suggested safe mechanisms. Every block holds a timestamp that encompasses preceding blocks' hash value, which keeps all transactions in a sequence order, thus ensuring that it is impossible to inject false transactions in the Blockchains. Furthermore, timestamps may be dispersed, eliminating node failure or data loss.

Randomly chosen collaborative nodes rerun the distributed IoT data [22]. It is worth noting that the number of cooperative nodes chosen at random has changed. To achieve consensus, each node does not need to solve the hash problem. While obtaining information, each node just seeks to assess the majority. The computational complexity of the IoT node can be minimized by incorporating a signature concept along the encryption/decryption performed in each node of the Blockchain. When using inherent function of large edge nodes, significant proportion verification is quite reliable. After storing the source data, the cloud service provider will create a new block by selecting one IoT edge node at random. If that was the first block, the hash of the previous block is set to zero. Each node updates the Blockchain by comparing its individual data with the publicly available information in Blockchain. Once the data stored in the blocks are transmitted to the cloud server space, each block node is emptied to save the storage capacity of each edge node of IoT.

The interior users/devices use a decentralized Blockchain [23] based authorization entrustment and accessing IoT control systems that incorporate a hierarchy of micro and macro consensus mechanism. Consequently, by allowing external users to get authentication to its parent IoT domains, the privacy of external users can be protected. The mechanism of proving the authenticity or integrity during authentication of user utilizes the hash values in describing or repossessing the user application, and those values are stockpiled along the localized Blockchain. Blockchain authorizes the IoT systems after effective authentication based on the validation of delegation procedures stored on Blockchain. An intruder uses a single or several compromised nodes to overpower the targeted node and generate denial-of-service attacks. The availability of the target system is obliterated by these attacks. Assume a hacker has installed malware on a user's computer or an IoT system. The intruder now intends to use the compromised user's delegation procedure to launch a DoS or DDoS attack on other users/IoT computers. Therefore, Blockchain-based authentication provides access to transaction on successful authentication.

3.3.2 Zero-Knowledge Proof

One of the most interesting concepts in applied cryptography currently is the zero-knowledge proof. A zero-knowledge proof is a powerful example of cryptographic creativity, with applications ranging from unilateral proliferation to providing anonymous and stable transactions for public Blockchain networks. Although Blockchain is a reliable option for reaching the required privacy and protection in exchanging information, it has one drawback: transparency. As a result, zero-knowledge proof (ZKP) has become one of the most common methods for meeting transaction confidentiality requirements. ZKP is a procedure whereby the prover (P) will send a proof to the verifier (V) stating that he is aware of secret value without revealing other details. ZKF is modest in demonstrating declaring someone knows something by merely disclosing it. The task is to defend such ownership without disclosing information or any other details.

The three primary requirements must be met by a ZKF:

1. Completeness: If the statement is valid, an honest prover will persuade the honest verifier—the one who is following the protocol correctly—of this reality.

2. Reliability: Except for a slight chance, no cheating prover can persuade the honest verifier that the assertion is true if it is not so.

3. Zero knowledge: If the statement is valid, the only thing the verifier knows is that the statement is true. To put it another way, understanding the declaration (but not the secret) is enough to conjure up a scenario demonstrating that the prover is aware of the secret. This is accomplished by providing a simulation to each verifier, which can generate a statement that appears similarly truthful to both P and V. The simulation model should be able to generate the transcript, although only having access to the prover's argument.

A ZKP has a general structure that comprises three sequential acts among participants A (prover) and B (verifier). A *witness*, a *challenge*, and a *response* are the terms used to describe these acts. Figure 3.6 represents the general structure of noninteractive ZKP and interactive ZKP.

Witness: The fact that A knows the secret specifies a set of questions that A can always accurately answer. A begins by picking a question from the set at random and calculating a proof. The proof is then sent to B by A.

Challenge: B then selects a question from the collection and asks A to respond.

Response: A figures out the answer and returns it to B.

Unlike a standard zero-knowledge proof, a noninteractive ZKP's general structure consists of only one action between participants P and V, which is a witness. P transfers the hidden knowledge to a special function that creates a proof as a statement. The result is indeed a proof value of some kind. Following that, P transfers proof to V.

On a Blockchain, it is often important to have some anonymity—making transactions without revealing the client's identity or transactions that are not connected. It ought to be feasible for a client to conduct various transactions while maintaining anonymity. A ZKP is a strong cryptographic tool, and its application in Blockchain appears encouraging in cases where current Blockchain technologies will adapt a ZKP to meet unique business requirements centered on data privacy. Although Blockchain is a reliable

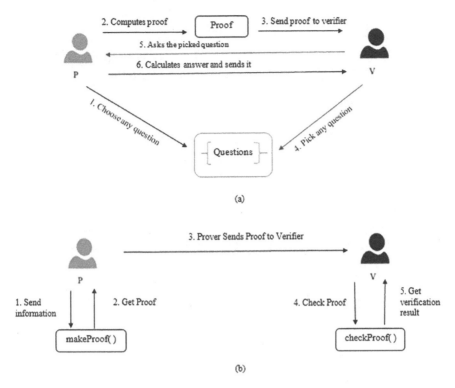

FIGURE 3.6 (a) Interactive ZKP; (b) noninteractive ZKP structure.

option for obtaining the targeted privacy and protection in exchanging information, it has one drawback: transparency. As a result, zero knowledge proof (ZKP) has been regarded as one of the most popular solutions for meeting transaction confidentiality requirements.

The scalability of ZKP will enable Blockchain to develop while also providing users mostly with expectation of flexibility, as well as improved control and freedom over their data. As a result, it is natural to speculate on how ZKPs can alter the Blockchain landscape. The following are the different areas where the zero-knowledge protocol is applicable in Blockchain.

1. Authentication Process

   With massively evolving information systems and related functionalities, the transmission of confidential information has gradually become a significant priority for businesses. With the assurance of enhanced protection, zero-knowledge proof will provide the ideal forum for the transmission of sensitive information such as authentication information. Zero-knowledge proofs can aid in the creation of a safe channel that allows users to use sensitive data without fear of being exposed. As a result, ZKPs will significantly aid in preventing data leakage problems in worst-case situations.

2. Messaging

The most common-use case for ZKPs in the Blockchain is messaging. In today's privacy-conscious world, end-to-end encryption is a must. The aim of message encryption is to ensure that no third party can see the correspondence between two people. Users must check their identity to the server, and the server must remember the user, according to most messaging platforms. Zero awareness evidence, on the other hand, enables the establishment of end-to-end trust in the messaging environment without requiring any additional details. As a result, the ZKP applications in messaging apps are the most important Blockchain applications.

3. Storage protection

The ZKP Blockchain link is also critical for resolving a major issue in the storage utility domain. ZKPs are thought to be one of the most useful tools for improving the performance of storage security methods and approaches. Along with the information in the storage unit, zero-knowledge proofs provide a procedure for storage unit protection. Above all, the access channels are protected by formidable protections, ensuring a secure storage and stable environment.

4. Using Blockchain to Submit Private Transactions

Among the various notable implementations of ZKP in Blockchain, one of the most influential is the transmission of private Blockchain transactions. When sending private Blockchain transactions, they must be protected from third parties. The most pressing issue in transmitting private Blockchain transactions is the numerous flaws that can be found in traditional methods. When zero-knowledge proof is combined with private Blockchain transactions, they become resistant to any kind of breaching or detection.

5. Validity of Blockchain Transactions Optimization

The case of ING is another well-known example of ZKP Blockchain implementations in the real world. To ensure transaction legitimacy,

ING has announced a solution that combines ZKPs with Corda, a well-known enterprise Blockchain platform. Corda's privacy and protection options were restricted until the launch of ZKPs. However, it now combines the two functions.

3.4 CURRENT CHALLENGES IN BLOCKCHAIN-BASED IoT APPLICATIONS

At present, 4G or 5G wireless communications face a number of obstacles. Meanwhile, the production of applications related to Blockchain-based IoT (B-IoT) are dynamic method influenced. As a result of various consistent features, integrating Blockchain into the combination adds more organizational and technological requirements. By specifying a Blockchain-based multilevel structure that specifies features, permission lists, and access privileges, access management to IoT networks can be made easier. It is worth noting, furthermore, that for several applications where anonymity is not needed, transaction security is still essential.

Unique identity authorization can remain an issue in the distributed IoT networks [24] where an identity authorizer is in charge of approving organizations; it can also be in charge of blocking them. To resolve this problem, a permissioned Blockchain can be used to protect as well as handle different IoT nodes. The emphasis is on resolving the privacy and reliability issues that arise from the use of centralized identification systems.

In order to ensure the protection of an information system, it must generally meet three criteria:

1. Confidentiality: The more confidential data should be safeguarded against unauthorized access.

2. Integrity: It ensures that unauthorized parties cannot modify or remove data. It is also common to add that if an approved party corrupts the data, the changes should be reversible.

3. Accessibility: Information can be processed whenever it is needed. In terms of reliability, Blockchain principles can be used to store and exchange confidential information in the cloud infrastructure or any service provider ensuring data integrity.

The increasing influence of Blockchain-based applications will help with the problem of establishing trust. Smart contracts based on Blockchain may be one of the core building blocks of future IoT trust infrastructures, as they are a requirement for business-critical interaction among devices without explicit human contact. Blockchain, on the other hand, necessitates computing resources and has a higher throughput bandwidth. It therefore restricts their use in IoT, necessitating the creation of new lightweight Blockchain-based solutions. Future research will focus on the effect of virtualization function of the network and the networks in which software is well defined on the authentication process in smart cities. The use of artificial intelligence (AI) for drone direction finding and movement within regions can be investigated further in order to boost the network's throughput.

Autonomous vehicles, highway units, and cloud/fog servers make up a smart, or intelligent, transportation system [7]. The internet allows these devices to connect with one another. It is a complex web of sensors, electronics, embedded systems, and other technologies that help navigate a complicated path. The system's intelligent units are responsible for making decisions quickly, consistently, and accurately. Travelers will enjoy a relaxed and secure ride in those variety of contact ecosystems. They are, indeed, susceptible to the various types of attacks. The use of Blockchain technology in quite an intelligent transport system strengthens and improves communications against external and internal security attacks.

Internet of Intelligent Things will be used to build a smart and intelligent healthcare system [25]. Smart healthcare devices are used in such contact environments. In order to prescribe medicine, health care professionals need the patient's health information, which must be communicated and accessed in a secure manner. A recommendation method that acts in the absence of a doctor may also be used in this correspondence.

Clarity of polls, voting administration, and elector registering and authentication are all handled separately in smart e-voting schemes in a dispersed setting using the Blockchain approach. In conventional schemes, the managing authorities are totally unaware of any criminal activity at a specific site. Using the Blockchain technology, all transactions are transparent to all election bodies at all levels. Additionally, electors are informed of the status of their ballots, which increases public confidence and strengthens democratic systems.

3.5 CONCLUSION

This study looked at the current state of Blockchain technology and suggested major scenarios for IoT implementations in areas like healthcare, security, smart cities, and resource management. In many ways, such as energy consumption in resource-constrained devices, these BIoT solutions face particular technological requirements that vary from deployments involving cryptocurrencies. The objectives of this review was to assess the functional limitations and identify potential research areas. It also offered a comprehensive approach, including detailed examination of enhanced Blockchain-based system design facets, such as its structural design using cryptosystem approaches and agreement schemes. Hence, it can be concluded that the IoT is still in its progression stages, and that wider adoption would necessitate further technical research advances to meet unique demands, as well as cooperation between organizations and governments.

REFERENCES

[1] Rathee, Geetanjali, Razi Iqbal, Omer Waqar, and Ali Kashif Bashir. "On the design and implementation of a blockchain enabled E-voting application within IoT-oriented smart cities." *IEEE Access* 9 (2021): 34165–34176.

[2] Fraga-Lamas, Paula, and Tiago M. Fernández-Caramés. "A review on blockchain technologies for an advanced and cyber-resilient automotive industry." *IEEE Access* 7 (2019): 17578–17598.

[3] Das, Ashok Kumar, Mohammad Wazid, Neeraj Kumar, Athanasios V. Vasilakos, and Joel J. P. C. Rodrigues. "Biometrics-based privacy-preserving user authentication scheme for cloud-based industrial Internet of Things deployment." *IEEE Internet of Things Journal* 5, no. 6 (2018): 4900–4913.

[4] Ding, Sheng, Jin Cao, Chen Li, Kai Fan, and Hui Li. "A novel attribute-based access control scheme using blockchain for IoT." *IEEE Access* 7 (2019): 38431–38441.

[5] Mandal, Shobhan, Basudeb Bera, Anil Kumar Sutrala, Ashok Kumar Das, Kim-Kwang Raymond Choo, and Youngho Park. "Certificateless-signcryption-based three-factor user access control scheme for IoT environment." *IEEE Internet of Things Journal* 7, no. 4 (2020): 3184–3197.

[6] Yao, Yingying, Xiaolin Chang, Jelena Mišić, Vojislav B. Mišić, and Lin Li. "BLA: Blockchain-assisted lightweight anonymous authentication for distributed vehicular fog services." *IEEE Internet of Things Journal* 6, no. 2 (2019): 3775–3784.

[7] Chen, Chien-Ming, Bin Xiang, Yining Liu, and King-Hang Wang. "A secure authentication protocol for internet of vehicles." *IEEE Access* 7 (2019): 12047–12057.

[8] Yazdinejad, Abbas, Reza M. Parizi, Ali Dehghantanha, Hadis Karimipour, Gautam Srivastava, and Mohammed Aledhari. "Enabling drones in the internet of things with decentralized blockchain-based security." *IEEE Internet of Things Journal* 8 (2020): 6406–6415.

[9] Cha, Shi-Cho, Jyun-Fu Chen, Chunhua Su, and Kuo-Hui Yeh. "A blockchain connected gateway for BLE-based devices in the internet of things." *IEEE Access* 6 (2018): 24639–24649.

[10] Hammi, Mohamed Tahar, Badis Hammi, Patrick Bellot, and Ahmed Serhrouchni. "Bubbles of Trust: A decentralized blockchain-based authentication system for IoT." *Computers & Security* 78 (2018): 126–142.

[11] Wang, King-Hang, Chien-Ming Chen, Weicheng Fang, and Tsu-Yang Wu. "A secure authentication scheme for internet of things." *Pervasive and Mobile Computing* 42 (2017): 15–26.

[12] Sharma, Pradip Kumar, Neeraj Kumar, and Jong Hyuk Park. "Blockchain technology toward Green IoT: Opportunities and challenges." *IEEE Network* 34, no. 4 (2020): 263–269.

[13] Xie, Lixia, Ying Ding, Hongyu Yang, and Xinmu Wang. "Blockchain-based secure and trustworthy internet of things in SDN-enabled 5G-VANETs." *IEEE Access* 7 (2019): 56656–56666.

[14] Irshad, Azeem, Muhammad Usman, Shehzad Ashraf Chaudhry, Husnain Naqvi, and Muhammad Shafiq. "A provably secure and efficient authenticated key agreement scheme for energy internet-based vehicle-to-grid technology framework." *IEEE Transactions on Industry Applications* 56, no. 4 (2020): 4425–4435.

[15] Chen, Yu-Jia, Li-Chun Wang, and Shu Wang. "Stochastic blockchain for IoT data integrity." *IEEE Transactions on Network Science and Engineering* 7, no. 1 (2018): 373–384.

[16] Fernández-Caramés, Tiago M., and Paula Fraga-Lamas. "A review on the use of blockchain for the internet of things." *IEEE Access* 6 (2018): 32979–33001.

[17] Cui, Zhihua, X. U. E. Fei, Shiqiang Zhang, Xingjuan Cai, Yang Cao, Wensheng Zhang, and Jinjun Chen. "A hybrid BlockChain-based identity authentication scheme for multi-WSN." *IEEE Transactions on Services Computing* 13, no. 2 (2020): 241–251.

[18] Wazid, Mohammad, Ashok Kumar Das, Sachin Shetty, and Minho Jo. "A tutorial and future research for building a blockchain-based secure communication scheme for Internet of intelligent things." *IEEE Access* 8 (2020): 88700–88716.

[19] Zhao, Quanyu, Siyi Chen, Zheli Liu, Thar Baker, and Yuan Zhang. "Blockchain-based privacy-preserving remote data integrity checking scheme for IoT information systems." *Information Processing & Management* 57, no. 6 (2020): 102355.

[20] Oktian, Yustus Eko, and Sang-Gon Lee. "BorderChain: Blockchain-based access control framework for the internet of things endpoint." *IEEE Access* 9(2020): 3592–3615.

[21] Ouaddah, Aafaf, Anas Abou Elkalam, and Abdellah AitOuahman. "FairAccess: A new Blockchain-based access control framework for the internet of things." *Security and Communication Networks* 9, no. 18 (2016): 5943–5964.

[22] Yavari, Mostafa, Masoumeh Safkhani, Saru Kumari, Sachin Kumar, and Chien-Ming Chen. "An improved blockchain-based authentication protocol

for IoT network management." *Security and Communication Networks* 2020 (2020): 1–16.

[23] Gauhar, Ali, Naveed Ahmad, Yue Cao, Shahzad Khan, Haitham Cruickshank, Ejaz Ali Qazi, and Azaz Ali. "xDBAuth: Blockchain based cross domain authentication and authorization framework for internet of things." *IEEE Access* 8 (2020): 58800–58816.

[24] Hassija, Vikas, Vinay Chamola, Vikas Saxena, Divyansh Jain, Pranav Goyal, and Biplab Sikdar. "A survey on IoT security: Application areas, security threats, and solution architectures." *IEEE Access* 7 (2019): 82721–82743.

[25] Dwivedi, Ashutosh Dhar, Gautam Srivastava, Shalini Dhar, and Rajani Singh. "A decentralized privacy-preserving healthcare blockchain for IoT." *Sensors* 19, no. 2 (2019): 326.

CHAPTER 4

Integration of Blockchain to IoT

Possibilities and Pitfalls

Vandana Reddy

CHRIST Deemed to be University, Bangalore, India

CONTENTS

4.1 INTRODUCTION TO IOT INTEGRATION WITH BLOCKCHAIN (IoTBC)

Smart infrastructures are increasingly common in developing nations, and they are powered by Internet of Things (IoT) [1]. These days, the IoT has drawn immense interest from scholars, scientists, and businesspeople because of its ability to offer inventive administrations across different

DOI: 10.1201/9781003188247-4

applications. IoT flawlessly interconnects heterogeneous gadgets and objects to make an actual organization in which detecting, handling, and correspondence procedures are naturally organized and overseen without human intervention. Distinctive organization advances include cyber-physical systems (CPS), machine-to-machine (M2M), and wireless sensor networks (WSNs). As modern technologies continue to upscale, there is a need to push the legacy systems, which are deeply integrated in numerous aspects of our daily living and industrial operations, toward the IoTBC. Among with most important concerns is security, specifically ensuring that the organization's networks and operations are protected from intrusions by unauthorized third parties. Blockchain (BC), initially applied in digital currencies, has risen to be a profoundly effective protective innovation for IoT applications. BC operates on the principle of a decentralized, carefully designed, and value-based data set that provides a safe method for processing and storing data across an enormous number of organization participants [2]. In current settings, huge amounts of information created from enormous number of IoT gadgets might bottleneck an IoT framework, bringing about low quality of administration (QoS). BC's shared engineering is viewed as a potential answer for issues of weak links and bottlenecks. The integration of BC into IoT can defeat the weak link and fill in as a satisfactory means to safely and efficiently stock as well as cycle IoT. BC evidently eliminates the requirement for trust among elements, which in turn eliminates the need for trusted middlemen, reducing operational times and costs. In a BC framework, members rely on the power of a mechanical instrument rather than utilizing the power of an association that can often prove deceitful. It also permits trust among free gatherings that do not consent and depend on solitary outsider trust. Consequently, the higher consistency of the product code, the more prominent confidence in the framework and the lower the need for confidence in that specialized framework's designers or administrators. For example, anyone can concentrate on the open Bitcoin convention [3]. Hence, BC innovation causes members to accept that nobody should be trusted, and nobody can profess to be a trusted party. Nonetheless, BC intricacy, including high processing expenses and deferrals, is a test in the combination of BC with IoTs that have limited power and processing capacities.

4.2 ARCHITECTURE OF IoTBC

4.2.1 Basics of IoT

The IoT joins actual objects with the virtual world. Smart gadgets and machines are thus connected with one another and the Internet. They obtain important data about their immediate environment, process it, and share it with others. The gadgets perform explicit assignments. A sensor, for instance, gauges the temperature outside, and the smart air-conditioning system connected to it reacts by adjusting the heating or cooling. All of that is done with zero human involvement. The most fundamental structure is shown in Figure 4.1. The three-layered model gives us the clarity in the model to depict the basic functionality of the IoT, but it is not sufficient in addressing the research component. Thus, there was a requirement for us to develop and enhance the three-layered structure to five layers so that the complete functionality of IoT with the research component could be easily addressed. The first layer in this regard is the perception layer that takes care of the lower-level operations and administration of the network.

The business layer deals with the entire IoT framework, including applications, business and benefit models, and clients' security.

The description of Figure 4.1 is given here. The perception layer in the model is the administrative leader that takes care of the lower arrangement. The perception layer in the model is the administrative layer that takes care of the lower-level arrangements of the devices. There are few protocols that help administer the devices that are part of the IOT. The transport layer is the second layer, which takes care of the communications within the network [4–6]. The transport layer protocols help transfer the messages from one device to another. The next layer of the processing layer takes care of the outgoing messages. All the messages sent by different devices are deciphered by the network in the processing layer. The next

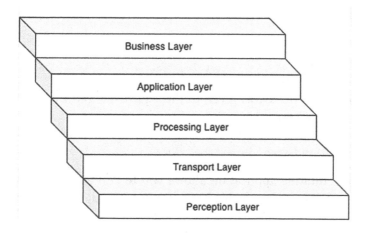

FIGURE 4.1   Layers of IoT.

layer provides a graphical user interface allowing to understand and send messages to one another.

4.2.2 Basics of Blockchain

BC is an organization of various gadgets (hubs) of similar significance associated with one another through the web. Basically, it records what has come in and gone out in a P2P format after the exchange has been confirmed by all interested hubs. This circulated record chips away at pre-characterized rules that are settled upon by every participating hub in the organization. These include a how-to for overseeing and approving exchanges, a calculation that characterizes the system for all partaking hubs to communicate with one another, and, occasionally, application programming interface. The guidelines that administer a BC network are called a convention. They represent the unwritten rules of the organization.

BC can be considered a variant of distributed ledger technology (DLT) [7] wherein interactions have been logged with an invariable cryptographic mark called a hash. The properties of DLT are [8]:

    i. Secure

    ii. Distributed

    iii. Anonymous

    iv. Programmable

    v. Immutable

    vi. Unanimous

    vii. Time Stamped

4.2.2.1 Layers of Blockchain

The five layers of BC are as given in the Figure 4.2 and the equivalent is depicted in the segment beneath:

1. Application Layer

   The facilitating application allows you to run all the decentralized applications accessing this layer. The facilitating convention will be completely decentralized also. Additionally, keeping up with these facilitating servers are totally secure. The decentralized applications

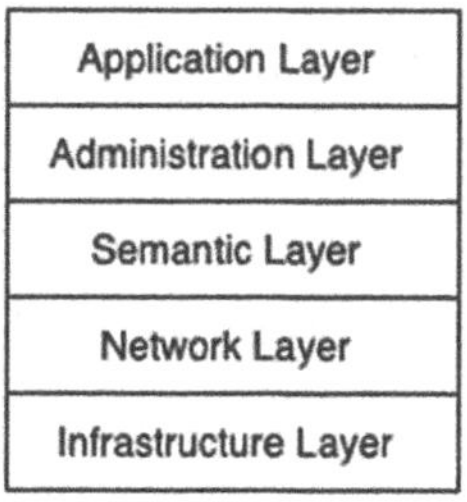

| Application Layer |
| Administration Layer |
| Semantic Layer |
| Network Layer |
| Infrastructure Layer |

FIGURE 4.2   Layer in blockchain.

are similar to today's application but with one unmistakable change: every one of them has a decentralized organization. Moreover, these are incredibly simple to make these days.

2. Administrations Layer

This is the second layer after the application layer, representing a cycle that assists with getting the most refreshed data from every one of the valid sources. Along these lines, it will assist the hubs with getting the most recent data about the network. On the other hand, off-tic registering is here to finish the figuring system outside the BC. It also provides advances extra security and takes the consumer to the center of the organization framework. Furthermore, designers get an administration structure here as well. Essentially this is a human-free, independent association that can create reasonable operational conditions. The genuine state channel is the pathway between two hubs. In this way, utilizing state channels, two hubs can speak with one another. Other than these, there are likewise different components in BC clarified layers. Predominantly these are oracles, multi-marks, smart agreements, digital assets, wallets, distributed record stockpiles, digital characters, and so on [9]. These are discretionary, since a BC innovation may or may not have them.

Oracles:

Oracles are important for tech-savvy smart contracts, since they go about as a specialist for gathering data from outside the organization.

Multi-signatures:

This component guarantees an alternate sort of safety convention. Essentially, you would have to sign any exchange without an extraordinary mark for making an exchange. Furthermore, here you can pick the number of these marks you need for executing.

Smart Contracts:

These are primarily self-executing legal agreements between two members on the BC innovation organization. As a rule, the entire framework disposes of the trust issue and allows you rapidly to trade any sort of resource. We will get to it later in the BC guide [10].

Digital Assets:

Presently on the BC innovation stack, the computerized resource can allude to anything. Generally it can mean digital forms of money, offers, gold, or different sorts of archiving. Any computerized component with genuine qualities would be known as advanced resources [11].

Wallets:

In the BC, wallets are used to store all computerized resources within the organization [12].

Distributed File Storage:

In the clarification of BC innovation, circulated record stockpiles are really a server area where every bit of data is stored. Clearly you'll need verification for getting to them [13].

Digital Identity:

As a rule, digital identities are the personalities of the clients in the organization [14].

3. Semantic Layer

In this layer, there are agreement calculations, virtual machines, any sort of investment necessities, etc. There is no BC network without agreement calculations. Generally agreement calculations are important for keeping an understanding between every one of the hubs. They facilitate interactions in which every hub operates with a similar understanding over the data on the record. Besides, in the record, nobody can simply begin an exchange and get it added. Those records may not be straightforward too. In this way, to ensure that the data on the square are legitimate, every hub proceeds via a similar arrangement. We will discuss this more later in the BC clarified guide. Next come the investment prerequisites [15]. These are predominantly assist the organization with concluding who can join the framework and who cannot. In addition, this component is essentially for the private BC advancements. Then again, virtual machines offer security and execution environment for every one of the assignments in the organization. Generally, it is utilized in the smart agreement execution. Next comes the side chains where engineers can go to a separate isolated BC environment to foster decentralized applications without influencing the main organization.

4. Network Layer

This layer contains the trusted execution environment (TEE), which takes care of versatility issues. It not only assists the organization with addressing this issue but also makes it safer. It also assists with storage when the amounts of data become excessive. Typically, these conventions are for when a standard convention does not completely acclimate to the framework. It makes an interface to assist the clients with conveying the BC network. Finally, the block conveyance network is an organization framework that conveys web content on demand. Generally you can see it in the common web design [16].

5. Infrastructure Layer

This is the last layer in the BC innovation engineering. In this one you may go over mining as an assistance convention. However, mining is gradually disappearing due to its extreme power requirements. Virtualization is the method for making any sort of virtual assets like servers, organization, stockpiling, OS, and so on. Besides, it works on a three-level framework. Hubs are likewise a part of this layer. Any gadget associated with the organization is viewed as a hub. No new hubs mean no BC innovation. One more important component of this layer is the decentralized stockpiling of the organization, making it safer against intrusion than at any other time in recent memory. Generally you may see tokens on this layer too. Tokens assist with keeping up with the biological system and are a local resource in the organization [17].

4.2.3 Integration of the Technologies: IoTBC

As modern technologies upscale, there is a need to push the legacy systems, which have been integrated into almost every interactive and executive aspect of the modern world and cannot be easily replaced toward the IoTBC, to improve their security and effectiveness.

The integration of the technologies mainly focuses on the following points:

1. Bringing computation and data storage as close as possible so that latency is low and the bandwidth is saved.

2. The goal of IoTBC is to move physically the computation from the data centers to the edge networks.

3. Creating IoTBC for DTL.

This pushes us to create a new system that can easily embed itself into the legacy system. To embed the innovative technology on the legacy systems requires orchestration. This orchestration can make both the existing technologies and innovative technologies compatible. Figure 4.3 describes the requirement of orchestration and the need to bring the legacy network and future network closer.

As given in Figure 4.3, the legacy network could be considered an IoT network. From Figure 4.1 it is clear that the IoT consists of five layers, from the perception layer to the business layer. In the legacy network we can consider the sensors and devices constituting the layers from the preliminary to the perception. The BC system is similar in this regard. The preliminary layer is the infrastructure layer that is comprised of data

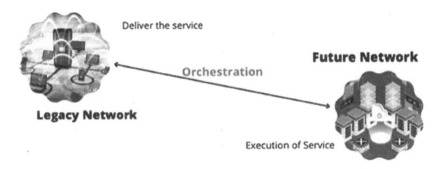

FIGURE 4.3 Merger of the legacy network and the future network via orchestration.

storage devices. Hence, in a future network made possible via orchestration, the preliminary layer would be the infrastructure layer. The second layer of our legacy network with respect to IoT is the transport layer; similarly, with respect to BC, it is a network layer. Both functionalities could be merged into the future network as a communication layer that provides the features of both networking and transportation. With this the services of communication layer could also be extended for providing trusted execution environment (TEE). Moving on further, the third layer in IoT is the processing layer, and with respect to BC, the third layer is the semantics layer. From the IoT perspective, the computational technology has been considerably improved in this layer. Figure 4.4 provides the detailed insight into the computational enhancements that have taken place in the third layer of IoT.

The processing in the third layer of IoT is a set of devices working together as a cluster with the data exchange from each device. Further, the devices that are made to communicate must be distributed because of the space and scalability constraints. Then the process moves to parallel computing. But these approaches include only the device, and the resources to be accessed form a different space. This is excessively time consuming, and thus the next method proposed, called grid computing, considers devices and resources together. The next constraint is on the computation that has to be moved to the common space of devices and resources made possible through cloud computing; the latest form of such processing is edge computing. For the future networks the integrated layer can become the computation layer, which serves as the third layer. With respect to BC, the third layer is responsible for recording the documents and smart contracts. If this layer is decentralized in the future network, then the

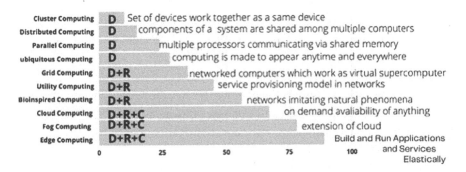

FIGURE 4.4   Computational enhancements in layer 3 of IoT.

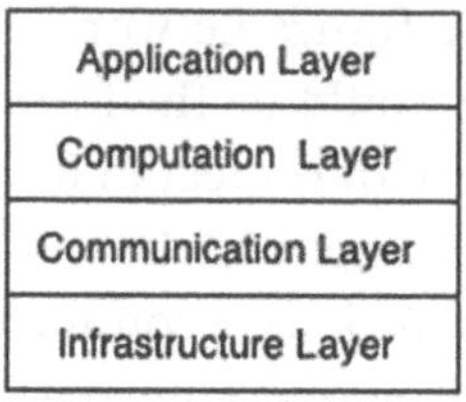

Application Layer
Computation Layer
Communication Layer
Infrastructure Layer

FIGURE 4.5   Layers of IoTBC.

computational overhead and security can be easily enhanced. Finally, the last layer of the future networks could be the application layer, which could merge layers four and five of IoT and BC and the cloud; this would be called the application layer. The new system thus would contain four layers rather than five, namely infrastructure, communication, computations, and application. Figure 4.5 presents the layers architecture of IoTBC representing the potential future network.

The merger of IoT and BC can be easily understood with respect to the orchestration given in the Figure 4.5. The example of IoT is taken to be a small smart voice calling system on a landline network. Onto this available legacy system we can embed the four-layered IoTBC, which is our future network.

As seen in Figure 4.6, a simple voice calling system could be made a smart voice calling system with the help of IoTBC. The legacy system for a simple voice calling system consists of the mobile stations connecting to the base stations. The way stations would in turn contact the base station controller, and the data services are provided by packet control unit. Later the mobile switching centers and gateway switching centers would coordinate the call and connect it to the nearby telephone network. The mobile switching centers would in turn contact a few support nodes and authentication centers to verify the call. Finally, the call, depending on the services, could also be connected to the internet. This legacy system could be made smart with the help of a simple orchestration.

Layer 1 in the orchestration system consists of the infrastructure that makes up the voice calling system. The infrastructure could be an amalgamation of the mobile phones and the base stations with respect to IoT, and a simple virtual machine can be integrated to the devices that are present here to incorporate BC layer 1.

Layer 2 in Figure 4.6 is the communication layer, which must include the networking and transportation functionalities of the IoT as well as data

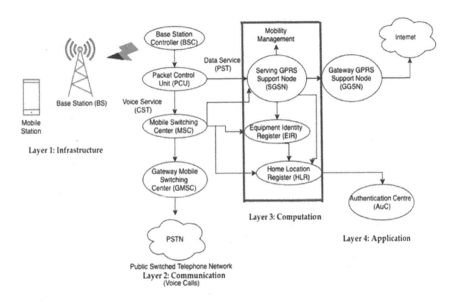

FIGURE 4.6   Orchestration on the simple voice calling system.

networking functionalities of the BC. Some of the networking protocols could be used in order to coordinate the infrastructure systems with respect to IoT. For the BC there must be the organization layer, otherwise called the P2P organization, which builds up correspondence between hubs. Sharding is a parting technique that appropriates processing and capacity jobs across a P2P organization to such an extent that, unlike ordinary BC, each hub is not responsible for dealing with the whole organizations exchanges load, but rather handles data identified with its parcel, or shard.

The layer 3 in Figure 4.6 is the computation layer that incorporates the data layer functionalities of BC. Assuming a solitary exchange is adjusted or changed, the Merkle tree root is likewise altered (Figure 4.7).

4.3 EXPLORING THE POSSIBILITIES FOR IoTBC

4.3.1 iGovernance

iGovernance is about the improvement of the public administration with regard to better effectiveness, local initiatives, versatile working, and more active involvement of the public. iGovernance is tied in with utilizing innovation to work with and support better preparation and implementation, changing the manner in which public services are delivered. This incorporates iGovernment productivity plans as well as versatile operations.

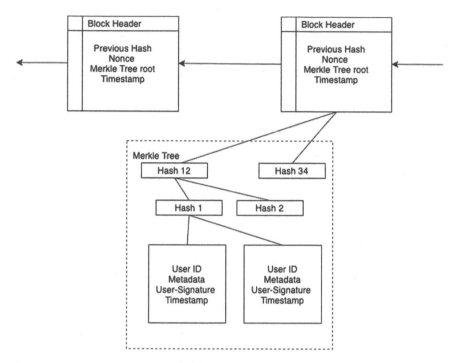

FIGURE 4.7 Basic structure of a block.

iGovernance includes such features as accountability, transparency, simplicity, and responsibility, which generally are viewed as the fundamental elements of civil service. The advanced bureaucracy is likewise liable for the detailing and execution of different projects and strategies of the public authority. Utilizing technology to oversee the service cycles will positively change managerial cycles generally [18].

Mechanization of Administrative Processes: This will limit human intercession, eventually eliminating any possibility of personal stake in the process of delivering a social service. This will likewise require organizational proficiency. Accessibility of online data for every office will empower web-based performance of tasks and document development to facilitate ease of planning, bookkeeping, and information stream control [19].

Documentation Reduction: It empowers exchanges of documents and other data for every representatives work area, improving development and utilization and reducing capacity burden. It is also likely to reduce waste by diminishing the volume of actual physical matter (i.e., paper) being used.

Quality of Service: E-data empowers associations as a compelling way to provide quality administrative services across the wide spectrum of individual requirements. This empowers civil servants to work more responsibly and effectively. Videoconferencing is one example of efficient distribution of services, allowing for a wider outreach and reducing the burden on service recipients to travel to a specific physical location to receive such services [20]. This also positively affects communal activities by opening them up to a wider circle of the public stakeholders.

iGovernance Guarantees: Achieves transparency via dissemination of information over the web, which includes point-by-point public review, making the delivery of services proficient and responsible.

Financial Development: iGovernance is more cost effective, both in highly populated areas as well as rural areas. In the former case it allows the streamlining of services and saving of resources, and in the latter it leads to an increase in information sharing among physically well-dispersed individuals within the same community, allowing for better resource utilization and improvements in quality of life across the entire spectrum of society [21].

Vital Data Provisioning: The often hierarchical nature of current public administration constrains people' capacities to accomplish data sharing and, in turn, innovations in delivery of public services. With iGovernance, this issue is alleviated.

4.3.2 iHealthcare

The significant parts of medical services frameworks are distinguishing proof, area, detecting, and network. iHealthcare is carried out through a wide scope of frameworks: crisis administrations, smart processing, sensors, lab on chips, remote checking, wearable gadgets, availability gadgets, and massive information.

The IoT-based frameworks are furnished with body sensor networks inside telemedicine frameworks. They incorporate gadgets with unique kinds of hubs that sense occasional distinction of patient information; to check the ventilation conditions for the patients in rooms, sensors are utilized to gather information for various measures adding to ventilation interaction of a room.

These sensors are customized to survey information of various reaches for temperature, tension, mugginess, and other environmental factors. These game plans help screen the patient conditions from a distance.

The framework can send intermittent reports to the medical clinic and keep up with the patient history. The clinic staff can see the information and set up the treatment plan for the patient under observation. Another type of gadgets utilized in IoT iHealthcare frameworks depends on remote sensor organizations [22].

Another, more complex aspect of iHealthcare involvement is remote medicine. In certain circumstances, IoT is the most dependable and least expensive arrangement, and the connection between various gadgets and intelligent correspondence frameworks likewise should be explored with more conventional destinations. Innovation makes it more straightforward to screen the patient well-being by sending data to iHealthcare groups like a specialist, attendants, and experts through IoT. It would be useful for experts to accumulate patient information utilizing store-and-forward strategy with the goal that it is available on demand.

Patient's current status, i.e., heart rate, temperature, position of body, blood glucose, and ECG, can be determined by utilizing sensors. The sensors are joined to Arduino UNO sensors that, when appended to the body Arduino board, get data and communicate them to the server. From this server the data are sent to the specialist who determines services, such as the need for medication. The concept of smart e-medicine includes real-time updates to individual electronic record, smart home administration, and smart clinical gadgets. While the nature of the marketplace is such that these advances will most likely appear first in developed countries, they are of a particular value to developing and less developed regions where individuals' access to medical services may be limited by large distances and the dearth of medical professionals [23].

4.3.3 iInfrastructure

Networking is essential for everyday activities. A framework can screen, measure, break down, convey, and act depending on information gathered by sensors. These foundations are put together not just with respect to their actual design (cabling, sensors, and so on) but also on four standards: information, investigation, criticism, and versatility.

Information: It is the essential component needed by a smart framework to work, and the natural substance needed by a smart foundation for its activity.

Examination: The investigation of data is vital to get helpful data for navigation.

Input: An information criticism circle is essential for any smart framework. This input is clear when data are gathered as to the manner by which a resource is utilized, and these data are utilized to further develop the manner in which the framework operates.

Flexibility: Smart frameworks are those that adjust to current requests, yet in addition adjust to the necessities of things to come.

Semi-Smart Foundation: This foundation gathers and registers information about its own utilization, its underlying conduct and environmental conditions, yet it has no capacity to settle on choices dependent on the acquired data. Examples of such a foundation would be maps that record a city's patterns of traffic congestion [24–26].

Intelligent Framework: This is a reference to frameworks that gather information to process and present the data that help a human with simple decisions. One illustration of this construction would be the traffic framework that recognizes heavy traffic areas and informs drivers so they can settle on better choices while they drive.

Smart Framework: This framework gathers information, processes data, and makes a suitable move totally independently (without human intercession), and progressively adjusts to evolving conditions. Current examples are smart organizations, structures, public frameworks, or seashores.

Smart Organizations: A smart organization is one that can effectively combine the conduct and activities of all clients associated with it so as to reduce risk and optimize operation. One model would be a high-voltage energy transmission network that is more mechanized and incorporated and better able to deal with every one of the gadgets associated with it because of its smart organization, accordingly accomplishing a productive and maintainable dispersion [27, 28].

Smart Structures: Smart structures are those that are operated using a high-level establishment and mechanical systems, an innovation that takes into account the mechanization of numerous inner processes, like heating, ventilation, lighting, and security, and different frameworks in the structure [29].

Benefits of iInfrastructures are:

Poise and Accuracy in Navigation: These frameworks consequently control the inner construction and the resource situation, monitor the environmental conditions, and utilize that to optimize their performance in terms of accuracy and speed of dynamic interactions between systems. One feature of a smart city, for example, would be knowing traffic conditions, public transportation patterns, and availability of parking spots to create and disseminate better routes for public and private vehicles on the road in a particular area.

Cost Efficiency and Investment Funds: These smart frameworks permit associations to exploit assets to accomplish more with less. One model could be controlling the utilization of power or heating in a structure.

Dependability: This alludes to diminishing downtime and unanticipated issues, permitting associations to keep offering their types of assistance with the highest and most consistent quality.

Well-being and Versatility: This suggests keeping up with cycles and plans that adjust to changes, guarding the framework and clients, and being resilient to human mistakes or cataclysmic events. An illustration of this could be the control of seating arrangements for a mass gathering.

Client Communication and Strengthening: Smart foundations improve the client experience and offer types of assistance that adjust to the changing requirements of purchasers.

Supportability: This benefit alludes to the reality of upgrading the dynamic interaction to guarantee a reasonable utilization of all overseen assets. An example is an effective waste, water, or energy board in an urban area.

4.4 PITFALLS OF IoTBC

The pitfalls of a BC framework are of the two types: permissionless and permission chains. Permissionless BC allows any party to take an interest in the organization, without verifying any bona fides, while permissioned BC chains are shaped by consortiums or a chairman who assess the support of an element in the BC system.

Regardless of the BC type, the business rationale is encoded utilizing smart contracts.

Blockchain dangers can be categorized as follows:

1. Standard Dangers: BC innovations open organizations to risks related to current business processes but with specifics for which entities need to account.

2. Esteem Move Hazards: BC empowers shared exchange of significant worth without the requirement for a focal intermediary. The worth moved could be resources, character, or data. The new configuration uncovers the cooperating gatherings to new dangers that in the past were overseen by focal intermediaries.

3. Smart Contracts Hazards: Smart agreements might conceivably encode complex business, monetary, and legitimate plans on the BC, and could bring about the danger related to the coordinated planning of these courses of action from the physical to the computerized systems.

The pitfalls of the IoT BC are as follows:

1. They are completely decentralized: The different in degrees of decentralization can be seen when you analyze public and private BCs. Public BC like Bitcoin or Ethereum are totally decentralized and open. Anybody can take an interest by adding or confirming information. A private BC permits only specific entities to take an interest in an organization. Each member is known, and each is allowed explicit privileges and limitations in the organization. This concentrates the BC more on the grounds that a more modest gathering controls the organization. Those with full access have a more concentrated authority than those with restricted admittance. The best-known private BCs are Ripple and Hyperledger.

2. They allow no data alteration: They are frequently commended for its security and straightforwardness. However, BC has a significant protection blemish in it, specifically the total inability of any user to change even their own information. Any data a user offers as part of the public exchange is no longer alterable. In other words, if you, after some period of time, encounter the circumstances in which the previously shared information has now become sensitive or damaging to you in any way, you have no way to delete it from the chain. On a public BC, everybody has a similar degree of access. This implies anyone can peruse the data. This means that the information you previously shared but now are unwilling to share still remains out there for anyone to access and to share up and down the chain. While private BCs attempt to resolve this issue by allowing each member a limited measure of authorizations, public BCs do not offer such an option.

3. They are not very efficient: It may be assumed that a technical innovation like BC is as quick in terms of processing as the existing systems, but this is in fact not so. In a framework that is fully decentralized, owned by no one, and run by clients, exchange confirmation does

not happen automatically but instead must be confirmed by individual clients. Checking an exchange to secure such a confirmation turns out to be slow. Blockchains today can deal with only around 7–15 exchanges per second. Non-Blockchain frameworks, even the legacy ones, can handle several thousand exchanges per second. As with any other drawback, there are efforts under way to resolve this. Moving to a more incorporated Blockchain can accelerate the cycle. This hinges on the fact that in such a framework fewer clients are expected to approve an exchange. Another technique can be moving to a proof-of-stake approval strategy. This approval technique gives those with more stake in the Blockchain higher approval capacity.

4.5 CONCLUSION

This chapter has focused on the basics of IoT and BC. The layered architecture and integration with layers have been described. The introduction section has presented the requirement and need of the hour for the integration. The sections that followed have described the methods, architecture, and generic frameworks where IoTBC can be applied.

REFERENCES

[1] Anusha Vangala, Ashok Kumar Das, Neeraj Kumar, and Mamoun Alazab. Smart secure sensing for IoT-based agriculture: Blockchain perspective. *IEEE Sensors Journal*, 21(16):17591–17607, 2020.

[2] Md Ashraf Uddin, Andrew Stranieri, Iqbal Gondal, and Venki Balasubramanian. Rapid health data repository allocation using predictive machine learning. *Health Informatics Journal*, 26(4):3009–3036, 2020.

[3] Jiawen Kang, Zehui Xiong, Dusit Niyato, Dongdong Ye, DongIn Kim, and Jun Zhao. Toward secure blockchain-enabled internet of vehicles: Optimizing consensus management using reputation and contract theory. *IEEE Transactions on Vehicular Technology*, 68(3):2906–2920, 2019.

[4] Avelino F Zorzo, Henry C Nunes, Roben C Lunardi, Regio A Michelin, and Salil S Kanhere. Dependable IoT using blockchain-based technology. In *2018 Eighth Latin-American Symposium on Dependable Computing (LADC)*, pages1–9. IEEE, 2018.

[5] Vivek Acharya, Anand Eswararao Yerrapati, and Nimesh Prakash. *Oracle blockchain quick start guide: A practical approach to implementing blockchain in your enterprise*. Packt Publishing Ltd, 2019.

[6] Licheng Wang, Xiaoying Shen, Jing Li, Jun Shao, and Yixian Yang. Cryptographic primitives in blockchains. *Journal of Network and Computer Applications*, 127:43–58, 2019.

[7] Caixiang Fan, Sara Ghaemi, Hamzeh Khazaei, and Petr Musilek. Performance evaluation of blockchain systems: A systematic survey. *IEEE Access*, 8: 126927–126950, 2020.

[8] Rahul P. Naik and Nicolas T. Courtois. *Optimising the sha256 hashing algorithm for faster and more efficient bitcoin mining*. MSc Information Security Department of Computer Science UCL, pages 1–65, 2013.

[9] Dan Boneh. *AggregateSignatures*, pages 27–27. SpringerUS, Boston,MA, 2011.

[10] Jae Cha Choon and Jung Hee Cheon. Anidentity-based signature from gap diffie-hellmangroups. In *International Workshop on Public Key Cryptography*, pages 18–30. Springer, 2003.

[11] Weidong Fang, Wei Chen, Wuxiong Zhang, Jun Pei, Weiwei Gao, and Guohui Wang. Digital signature scheme for information non- repudiation in blockchain: A state of the art review. *EURASIP Journal on Wireless Communications and Networking*, 2020(1):1–15, 2020.

[12] Avi Asayag, Gad Cohen, Ido Grayevsky, Maya Leshkowitz, Ori Rottenstreich, Ronen Tamari, and David Yakira. Helix: A scalable and fair consensus algorithm. Technical Report. Orbs Research, 2018.

[13] Ke Huang, Xiaosong Zhang, Yi Mu, Fatemeh Rezaeibagha, and Xiaojiang Du. Scalable and redactable block chain with update and anonymity. *Information Sciences*, 546:25–41.

[14] Md Ashraf Uddin, Andrew Stranieri, Iqbal Gondal, and Venki Balasubramanian. Blockchain leveraged decentralized IoTehealth framework. *Internet of Things*, 9:100159, 2020.

[15] Yu Zuoxia, Man Ho Au, Jiangshan Yu, Rupeng Yang, Qiuliang Xu, and Wang FatLau. New empirical traceability analysis of cryptonote-style blockchains. In *International Conference on Financial Cryptography and Data Security*, pages 133–149. Springer, 2019.

[16] Rebekah Mercer. Privacy on the block chain: Uniquering signatures. arXivpreprintarXiv:1612.01188, 2016.

[17] R. Yap. Understanding how zero coin in zcoin works and how it compares to other anonymity solutions part1, 2017.

[18] Chaoyang Li, Yuan Tian, Xiubo Chen, and Jian Li. An efficient anti-quantum lattice-based blind signature for blockchain-enabled systems. *Information Sciences*, 546:253–264, 2020.

[19] Jon M. Peha and Ildar M. Khamitov. Paycash: A secure efficient internet payment system. *Electronic Commerce Research and Applications*, 3(4):381–388, 2004.

[20] Ahsan Manzoor, An Braeken, Salil S Kanhere, Mika Ylianttila, and Madhsanka Liyanage. Proxyre-encryption enabled secure and anonymous IoT data sharing platform based on blockchain. *Journal of Network and Computer Applications*, 176:102917, 2020.

[21] Abdul Rahman Taleb and Damien Vergnaud. Speeding-up verification of digital signatures. *Journal of Computer and System Sciences*, 116:22–39, 2020.

[22] Leila Ismail and Huned Materwala. A review of blockchain architecture and consensus protocols: Use cases, challenges, and solutions. *Symmetry*, 11(10):1198, 2019.

[23] Vincent Gramoli. From blockchain consensus back to Byzantine consensus. *Future Generation Computer Systems*, 107:760–769, 2017

[24] Li Peng, Wei Feng, Zheng Yan, Yafeng Li, Xiaokang Zhou, and Shohei Shimizu. Privacy preservation in permissionless blockchain: A survey. *Digital Communications and Networks*, 7(3):295–307, 2020.

[25] Sujit Biswas, Kashif Sharif, Fan Li, Sabita Maharjan, Saraju P. Mohanty, and Yu Wang. Pobt: Alight weight consensus algorithm for scalable IoT business blockchain. *IEEE Internet of Things Journal*, 7(3):2343–2355, 2019.

[26] Saqib Hakak, Wazir Zada Khan, Gulshan Amin Gilkar, Muhammad Imran, and Nadra Guizani. Securing smart cities through blockchain technology: Architecture, requirements, and challenges. *IEEE Network*, 34(1):8–14, 2020.

[27] Fan Yang, Wei Zhou, Qing Qing Wu, Rui Long, Neal N. Xiong, and Meiqi Zhou. Delegated proof of stake with downgrade: A secure and efficient blockchain consensus algorithm with downgrade mechanism. *IEEE Access*, 7:118541–118555, 2019.

[28] Ahmad Firdaus, Mohd Faizal AbRazak, Ali Feizollah, Ibrahim Abaker Targio Hashem, Mohamad Hazim, and NorBadrul Anuar.Therise of "blockchain": Bibliometric analysis of blockchain study. *Scientometrics*, 120(3):1289–1331, 2019.

[29] Bin Yuan, Hai Jin, Deqing Zou, Laurence Tianruo Yang, and Shui Yu. A practical byzantine-based approach for faulty switch tolerance in software-defined networks. *IEEE Transactions on Network and Service Management*, 15(2):825–839, 2018.

CHAPTER 5

Design and Fabrication of IoT-Based Smart Home Automation System Using Wireless Fidelity Shield and Arduino Microcontroller

Ayan Bhattacharjee and Alak Roy
Tripura University (A Central University), Tripura, India

Mampi Devi
Tripura University (A Central University), Tripura, India

Sajal Kanta Das
Tripura University (A Central University), Tripura, India

CONTENTS

DOI: 10.1201/9781003188247-5

5.1 INTRODUCTION

Technology is responsible for making our lives better and easier, and it keeps changing as our needs change. One of the prominent examples of technology is a smartphone, which has become a practically inseparable component of our daily life [1]. Embedded systems have also gained immense popularity and have been used as key components behind the evolution of smart appliances. IoT is a gift when every device in our home and/or workplace is connected to the internet, making our daily activities easier to perform than ever before. At present, smart automated appliances and systems are making their presence felt in the technological market. Also, people are recently investing more to build smart homes and offices [2].

In addition to managing these systems, some companies are also trying to focus on Artificial Intelligent (AI)-controlled smart home automation systems, thereby releasing humans from direct interaction with their appliance es. However, the AI deployed in such systems at present is not able to do all the activities, and the cost of such a system is still very high [3]. Interacting with an existing smart home interface requires considerable technical know-how, however, which is clearly a significant obstacle for a person who does not possess such knowledge—which includes the majority of older persons, who have no "inherent" familiarity with modern electronics that the younger generations possess [4]. Therefore, the proposed system focuses not only on providing a low-cost smart home automation system but also on making it easy to use via an Android-based application.

This chapter proposes an IoT-based smart home automation system using a wireless fidelity (Wi-Fi) shield and an Arduino microcontroller. The aim of the proposed system is to provide users with an interface that allows controlling all the household electrical appliances (lights, fans, faucets, air conditioning) installed in homes with a relative ease of use and cost efficiently. To build the system requires an Android smartphone, a microcontroller (Arduino Uno R3), a Wi-Fi module (ESP-01 8266), a relay module, and a power supply responsible for powering the devices. The system is intended to be easily used by persons with only basic familiarity with modern electronic devices, of which older people represent the majority.

The chapter is organized as follows. Section 5.2 reviews the existing literature on the subject. Section 5.3 presents the proposed system with hardware and software requirements. Section 5.4 presents results and discussions. Finally, the chapter concludes with future research directions in Section 5.5.

5.2 LITERATURE SURVEY

This section highlights the recent literature in the field of smart home automation systems. In [1], the author proposed a low-cost home automation system built using DigilentchipKIT Uno32 and Arduino Uno microcontroller board and highlighted how people can easily monitor and know about the appliances being used at their homes by remotely accessing the system.

In [2], authors highlighted the energy consumption done by household appliances, and the impact of high energy consumption on natural surroundings. In addition, their paper focused on how it is possible to reduce the energy consumption based on parameters measured by various sensors plugged into our homes.

In [3], authors described how an automation system can exclude the need of human interaction with the various system, how using ATmega 328 and an Android application/SMS, a home automation system can be built for controlling electrical appliances installed in an office or home environment.

The authors of [4] focused on how people want to control their home appliances remotely and proposed an automation system built using 8051 microcontrollers, which could be controlled by SMS sent via phone using GSM technology.

In [5], the authors wrote about the impact of high energy consumption by all the electrical appliances used in people's home, and how due to the lack of access, these appliances lead to energy loss. The authors encouraged the use of smart home automation using Bluetooth, ZigBee, or Wi-Fi technology, which would allow people to interact and control their household appliances remotely. This would, in turn, reduce energy consumption.

The authors in [6] explained how home automation has gained immense popularity in a short period and has improved the quality of life. They also mentioned how people can develope smart home automation by using microcontrollers, Android smartphones, and wireless technology (ZigBee, Wi-Fi, or Bluetooth). They described both advantages and drawbacks of such developments

In [7], the authors described how, by using the Internet of Things (IoT), human direct interaction with the appliances could be reduced by placing various sensors and integrated systems that can not only sense but also fetch data to the internet. The resulting system would be not only more efficient and cost effective but also more secure.

The author of [8] discussed about their home automation system built using Raspberry Pi and mobile devices. Beside their system working, the author also described Raspberry Pi and its built-in support for several components to handle several appliances simultaneously.

In [9] the authors presented a home automation system as a type of a web-based application that can be supported by mobile devices for controlling domestic appliances installed on users' home, as well as how such a system can help conserve energy and provide security. The authors also highlighted the importance of proper security once all the appliances had automation support.

The authors of [10] discussed the existing smart home systems alongside the various multimedia devices and sensors integrated when building such a system. Their study also provided a guideline for future researchers on building a sustainable smart home.

In [11], the authors illustrated how to build a smart home automation using Raspberry Pi with a centralized web-based control using the IoT. The main idea is to have centralized control of all the domestic appliances and how such a smart system helped in building a secured home.

5.3 PROPOSED SYSTEM

This section presents the design and fabrication of the proposed IoT-based smart home automation system using a wireless fidelity shield. In addition, various hardware and software components required for building the system are presented.

5.3.1 Hardware Requirements

The hardware components required for the proposed systems are Arduino Uno R3 (microcontroller) and REES52 Optocoupler (relay module). The system also requires ESP-01 8266 (Wireless Fidelity shield) for establishing communication with an Android smartphone. The following contains the details of the various hardware components required for building the smart home automation system.

Arduino Uno R3: The Arduino Uno R3 is a microcontroller designed by Arduino. It is an open-source microcontroller board and is the most-recommended board for getting started with this type of a smart home system. Even though Arduino is the official maker of the board, other manufacturers also build this board and sell it at a lower price than the original maker. The Uno board comes with a USB port, which is used for connecting with a computer and for uploading coding (sketches) from the Arduino IDE. In the hardware part of the proposed system, Arduino plays a crucial role, as it is responsible for powering up not only the ESP-01 8266 Wi-Fi module but also the relay module. Moreover, the ESP-01 8266 lacks a direct communication port with a computer, hence a connection must be made via the Arduino UNO R3 for uploading the program in the ESP-01 8266.

ESP-01: The ESP-01 8266 Wi-Fi module is a low-cost microchip having full TCP/IP stack support and has the capability of a microcontroller. Generally used for connecting microcontrollers to the internet, this Wi-Fi module can be programmed to connect to a Wi-Fi access point or even to act like one. In this particular system, the ESP-01 8266 is programmed to connect with the available internet access point (router, smartphone), and once connected with an access point. One can control the appliances connected to the system in their homes. The ESP-01 requires 3.3 V to run, which is supplied to it from the Arduino Uno R3 board. As shown in Figure 5.1, the ESP-01's TX and RX pins are connected to Arduino's TX and RX for communicating with the computer. The general-purpose input output (GPIO) pins are connected to the input pins of the relay module.

REES52 Optocoupler: The last component needed is the REES52 Optocoupler relay unit module, which can be considered an on/off switch for various appliances, providing input usually by a microcontroller. Depending on the requirement, it is available with a two-, four-, or even eight-way relay unit. Similar to the ESP-01 8266, the

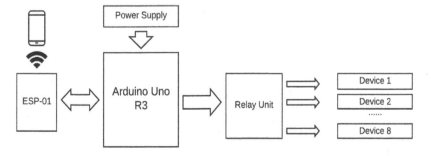

FIGURE 5.1   System design of the smart home automation system.

Relay unit also consumes power (5 V) supplied from the Arduino Uno board. Based on the input received from the ESP-01 8266 module, the relay unit closes or opens the circuit, thereby turning on or off the appliances connected to it.

5.3.2 Software Requirements

The software components required for the proposed systems are Arduino IDE and Android Studio that play the pivotal role in making the program and the Android app.

Arduino IDE: It is free open-source IDE used for making and uploading programs (sketches) to Arduino, ESP, and other microcontroller boards. Once it has completed writing the program, it is possible to verify it for checking errors and then upload it to the microcontroller board. Arduino IDE programming language is based on C++ programming language, so those familiar with C++ will be able to catch up with the Arduino IDE much faster.

In this project, Arduino IDE is required for uploading the code (sketch) to the ESP-01 8266 module. As a result, the ESP-01 8266 not only connects to the internet but also can receive input via its general-purpose input output (GPIO) pins. The proposed system features an Android app, which interacts with the ESP-01 8266 module for controlling the appliances.

Android Studio: Like Arduino IDE, Android Studio is also available for free through Google. The IDE is used for building professional Android applications. Android Studio comes with Android Emulator support, which lets the programmer run and deploy the app to find bugs and test the built application. By deploying an app in the emulator, the programmer can check whether the app is running or crashing while being used. Android Studio is used for building, testing, debugging, and deploying the app. Most importantly, the built app is fully customizable; therefore, it can be extended for controlling other kind of devices and showing real-time information to the users.

The core concept of the system relies on wireless communication, which allows a user to interact and control the home appliances remotely from any place they want. To fulfill the need, wireless technology (Wi-Fi) is used for enabling the microcontroller controlling the system to get connected to their internet, thus forming the communicating gateway with the Android smartphone for receiving and transferring the inputs and outputs. In this system, an Android application is required for controlling the system. The Android smartphone runs an app that interacts with the built system and controls the appliances connected to the automation system.

5.3.3 System Design

The system design for the IoT-based smart home automation system is presented in Figure 5.1, where ESP-01 module is connected to the relay unit via Arduino Uno R3 microcontroller. The output of the relay unit is connected to the electrical appliances. For making the connection between the multiple components of the system, jumper wires and bread-board were used. The ESP-01 and the relay unit require female jumper wires whereas the Arduino Uno R3 requires male jumper wires. Overall, a combination of jumper wires and breadboard are used for setting up the smart home automation system. The system needs a power source, which can be provided using a normal power bank. Once the system is turned on, it will look for a nearby access point and get connected using SSID and password. Once the device is connected, the user can start the Android application to interact with the system.

The wire layout for the proposed smart home automation system is shown in Figure 5.2. The connection between Arduino Uno R3 and ESP-01 8266, ESP-01 8266, and RESS52 relay module are given in Table 5.1.

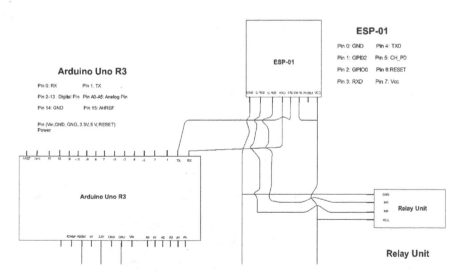

FIGURE 5.2   Wire layout of the smart home automation system.

TABLE 5.1 Pin Connection of the Smart Home Automation System

Arduino Uno R3	ESP-01 8266	RESS52 Relay Module
RX (Pin 0)	RX (Pin 4)	
TX (Pin 1)	TX (Pin 5)	
3.3 V	VCC (Pin 7), CH PD (Pin 5)	
5 V		VCC
GND	GND (Pin 0)	GND
	GPIO 0 (Pin 2)	INP 1
	GPIO 2 (Pin 1)	INP 2

5.4 RESULTS AND DISCUSSIONS

After testing, it was found that when both the smart phone and the smart home automation system are disconnected from the internet, the page fails to load. As a result, the user gets a popup informing that the user's device is not connected to the internet. On the other hand, when the smartphone and smart home automation system are connected to the internet, the app loads the interface of the home automation system featuring all the buttons. Based on the button pressed, the connected device powers up, and simultaneously the information is highlighted on the app. The layout to develop the designed proposed system is shown in Figure 5.3.

In addition, when the ESP-01 could not respond to the tapped button, it responded with delay. From the figure it is visible that the interface contains only two buttons. However, the application can be customized and more buttons added to the interface. Due to the fact the ESP-01 comes

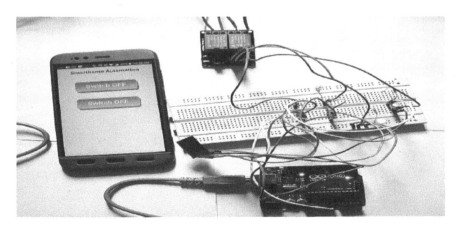

FIGURE 5.3 Output interface of the smart home automation system.

with just two GPIO pins, the buttons have been limited to two. Therefore, the number of switches depends on the model of the ESP module being used.

5.5 CONCLUSION AND FUTURE DIRECTION

The chapter proposed an IoT-based smart home automation system using a wireless fidelity shield and an Arduino microcontroller that can convert an ordinary home into a smart home. The system is customizable and thus can be extended by adding sensors, a monitor screen, and other related devices. The aim of the chapter is to show how to build a low-cost smart home automation system. The system demonstrated how IoT-based smart home automation can improve the standard of living. The proposed system can save energy and helps older and disabled people control their household appliances without physical movement using a simple, user-friendly interface on their smartphone. Moreover, the system installation process is a one-time process that does not require any modification. The only thing required is to set up the connection of the appliances.

The future work includes replacement of the existing ESP-01 and Arduino Uno R3 with ESP NodeMCU microcontroller and integrating sensors with the existing devices. The future work also includes adding security features with alert capabilities with the systems.

REFERENCES

[1] Stolojescu-Crisan, C., Crisan, C. and Butunoi, B.P., 2021. An IoT-based smart home automation system. *Sensors*, *21*(11), p. 3784.
[2] Satheeskanth, N., Marasinghe, S.D., Rathnayaka, R.M.L.M.P., Kunaraj, A. and Joy Mathavan, J., 2022. IoT-based integrated smart home automation system. In *Ubiquitous intelligent systems* (pp. 341–355). Springer, Singapore.
[3] Yuvaraj, K., 2021. Smart home automation system. *Annals of the Romanian Society for Cell Biology*, pp. 9087–9090.
[4] Ray, A.K. and Bagwari, A., 2020, April. IoT based smart home: Security aspects and security architecture. In *2020 IEEE 9th International Conference on Communication Systems and Network Technologies (CSNT)* (pp. 218–222). IEEE.
[5] Isyanto, H., Arifin, A.S. and Suryanegara, M., 2020, February. Design and implementation of IoT-based smart home voice commands for disabled people using google assistant. In *2020 International Conference on Smart Technology and Applications (ICoSTA)* (pp. 1–6). IEEE.
[6] Singh, H., Pallagani, V., Khandelwal, V. and Venkanna, U., 2018, March. IoT based smart home automation system using sensor node. In *2018 4th International Conference on Recent Advances in Information Technology (RAIT)* (pp. 1–5). IEEE.

[7] Mahamud, M.S., Zishan, M.S.R., Ahmad, S.I., Rahman, A.R., Hasan, M. and Rahman, M.L., 2019, January. Domicile-an IoT based smart home automation system. In *2019 International Conference on Robotics, Electrical and Signal Processing Techniques (ICREST)* (pp. 493–497). IEEE.

[8] Jabbar, W.A., Alsibai, M.H., Amran, N.S.S. and Mahayadin, S.K., 2018, June. Design and implementation of IoT-based automation system for smart home. In *2018 International Symposium on Networks, Computers and Communications (ISNCC)* (pp. 1–6). IEEE.

[9] Jabbar, W.A., Kian, T.K., Ramli, R.M., Zubir, S.N., Zamrizaman, N.S., Balfaqih, M., Shepelev, V. and Alharbi, S., 2019. Design and fabrication of smart home with internet of things enabled automation system. *IEEE Access, 7*, pp. 144059–144074.

[10] Agarwal, K., Agarwal, A. and Misra, G., 2019, December. Review and performance analysis on wireless smart home and home automation using IoT. In *2019 Third International Conference on I-SMAC (IoT in Social, Mobile, Analytics and Cloud) (I-SMAC)* (pp. 629–633). IEEE.

[11] Mahmud, S., Ahmed, S. and Shikder, K., 2019, January. A smart home automation and metering system using internet of things (IoT). In *2019 International Conference on Robotics, Electrical and Signal Processing Techniques (ICREST)* (pp. 451–454). IEEE.

CHAPTER 6

Block Chain Architecture in Financial System for Integrity, Transparency, and Trust-Free Transaction

Jitendra Saxena

The Institution of Engineers, Kolkata, India

CONTENTS

DOI: 10.1201/9781003188247-6

6.1 INTRODUCTION

Blockchain applications were initiated with Bitcoin in 2009, and it has survived since then, but in-depth research findings for Blockchain technologies are yet to be implemented and validated. Innovative findings and initiatives that have significant societal impact are still on the way, and every day is a new day with new sunrise and spectrum. The recent trends and development and decreasing consumer trust on banking sectors open the door for Blockchain technology to enter in the pathway of financial system that will be fully transparent and holistically adapted for societal growth and sustainability. Blockchains used to strengthen cryptocurrencies is one of a wide variety of applications that are used extensively in Blockchain architecture, and simultaneously an Ethereum foundation has started a new paradigm by implementation of generic programmable Blockchain that is globally adopted and implemented. The resulting outcomes establish the pathway of high-end features such as 100% transparent, trust-free, reliable, and highly secure nature for different emerging areas other than the financial systems. The goal of this chapter is to investigate and compare the focal points of currently trust and fully transparent centralized systems in the financial sector and how global society will benefit by adopting fully trust-free and transparent system.

Blockchain system innovators have the vision of big companies who have billion-dollar business transactions and have a huge volume of data from which one can extract more information. There are various types of issues about privacy and transparency [1, 2]. Users' behavior and dynamics and privacy concerns are the main findings. Many banks are smarter these days and have more information and control. Banks can track all types of information. There is a lot of data in bank databases, and they have more information controlled by the bank, and so there is no privacy. There are a lot of advantages and disadvantages in social and public media. Taking the example of land records of the government, we can easily manipulate or modify data so there is a lot of potentials to avoid manipulation and to apply high-technology framework for the development of society. Infrastructure to maintain data is also of prime importance, and a lot of manpower is required to maintain infrastructure, which should be optimized. As per business transactions, if the database is leaked, the trust of the organization will no longer exist. The vital issues are privacy and security of the data [3, 4]. Therefore, the concept of decentralization to maintain security and privacy of data is of paramount importance. The problem is centralization of transactions, and the solution is decentralization [5–10]. To track the evolution of technology: in 2000, web technology 1.0 was the prevailing one, in 2010 it was web 2.0, and after 2015 until now it is web 3.0.

6.2 FUNDAMENTALS OF BLOCKCHAIN TECHNOLOGY

For every technological system there is one web server with a database for scaling up the data or expanding the operation. A data system is usually limited in its storage and upgrade by the size (capacity) of the server it utilizes. In instances of sudden surges in data usage, it is often quite difficult to achieve a server upgrade in a timely manner. To solve this problem, an organization can rent extra server space without obtaining the actual physical server, like the one provided by Amazon Web Servers. All the data are stored in these rental servers. Decentralization of data and decentralized apps (D Apps) are the common practices in a Blockchain framework. Building a generic framework of a Blockchain architecture is key, and a fundamental issue with Etherium and hyper ledger includes the important financial transactions for Blockchain. Amazon, for example, with its billions of users, is a centralized network utilizing a system called grid computing. Figure 6.1 shows Amazon-based centralized system with users and different functionalities such as customer, types of customer requirement, different types of product information, data related to images and videos, and digital marketing. It is a centralized system with a generic processing unit for fast communication and infrastructure facilities.

Within a Blockchain architecture, the information is not owned by a single entity but rather by multiple individual users. In a centralized data transactions system there is a high chance of manipulation of data. In a decentralized financial transactions system, such chances are significantly reduced because there is no single identity controlling the transactions

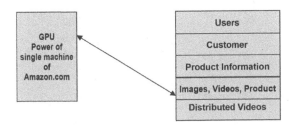

FIGURE 6.1   Centralized Amazon system with specific data features.

[11–15]. Data deletion and manipulation are avoided in a decentralized network; data are replicated for all users, and everybody receives the information. The identities of senders and receivers are not disclosed and are encrypted by a security and hash algorithm (SHA 256).

6.3 OVERVIEW AND IMPLEMENTATION OF BLOCKCHAIN IN DECENTRALIZED FINANCIAL NETWORK

The important features of a Blockchain network are:

1. Immutability

2. Privacy

3. Transparency

Figure 6.2 provides a detailed overview of converting a private bank system to a decentralized system in which shared ledger and records are encrypted by security and hash algorithm (SHA 256) and its transactions are converted into Blocks like B1, B2, B3, and B4.

Based on the overview and consensus of social media, people, and different types of users, transactions with cryptographic puzzles are solved and executed. The financial information and transactions are registered, verified, and validated by miners and a consensus algorithm [16–19].

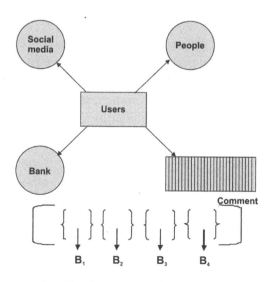

FIGURE 6.2   Decentralized bank system with private Blockchain architecture.

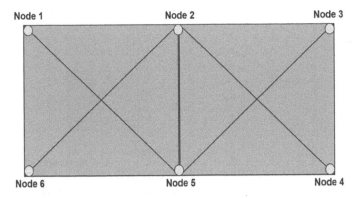

FIGURE 6.3   Decentralized mesh Blockchain architecture of nodes transactions.

Figure 6.3 shows a decentralized mesh Blockchain architecture of node transactions in which a network of nodes with their respective data and metadata are registered and validated and confirmed by the formation of a block.

6.4 BLOCKCHAIN ARCHITECTURE IN FINANCIAL SYSTEM

Figure 6.4 shows a block with a financial transaction, consisting of different functionalities of preprocessing of data with security and transparency

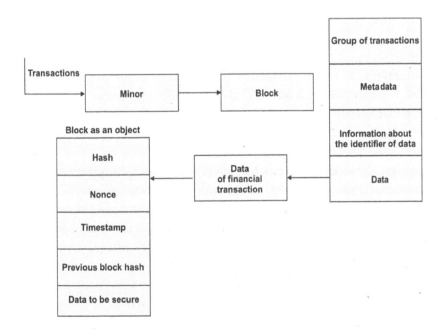

FIGURE 6.4   Financial transactions with minors and Blockchain architecture.

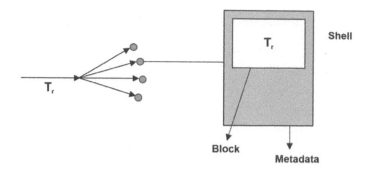

FIGURE 6.5   Formation of a block after execution of transactions with data and metadata.

of network. The miners with proof of work authenticate transactions with the creation of a block. A block consists of confirmed transactions after execution of specific features of hash, nonce, time stamp, and previous block hash, and ultimately data is secured with little chance of identification of data, resulting in transparent and secure transactions.

Figure 6.5 represents a financial transactions shell with data and metadata in an active block. These features are used in a real-time Blockchain network. We can see the real-time transaction in a Blockchain explorer. Miners give permission to maintain a ledger. Miners are the nodes or transaction points where financial transactions are received. Nodes are the miners in which financial transactions are done. A miner receives a transaction, which arrive in the form of a key, along with the value of the transactions. When a transaction is given to multiple people in the network, a miner creates a block. Only one miner can create a block at any particular instant.

6.4.1 Proof of Work

Figure 6.6 shows the proof of work solving cryptographic puzzles for financial transactions. A network server with its application interface sends a cryptographic puzzle. The proof of work is of paramount importance for creating logical puzzles and nonce. The solutions of the logical puzzles are fed to nonce, which acts as a miner to form a block after confirmation of the validation of the proof of work, which certifies the transaction.

Figure 6.7 shows the launching of a Blockchain architecture on a local host web server applications. The local host 3001 will be the first transactions node and will be registered first and then broadcast with other local host servers like 3002, 3003, etc.

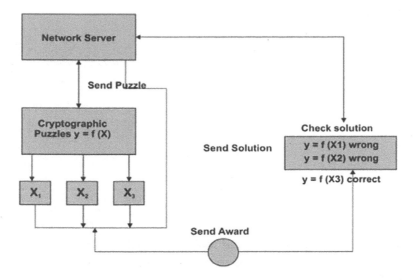

FIGURE 6.6   Proof of work solving cryptographic puzzles for financial transactions.

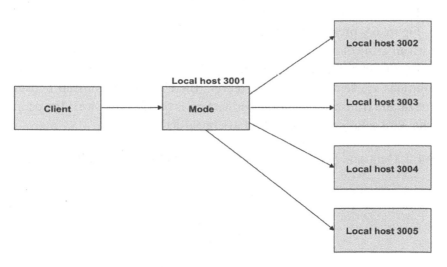

FIGURE 6.7   Launching of Blockchain architecture on local host web server applications.

The proof of work is the core functionality for the creation of a block. The back-end architecture of the proof of work is done by solving cryptographic puzzles.

Figure 6.8 provides a detailed architecture setup with a shared ledger for financial transactions to all nodes. A block inside the shared ledger consists of validated and encrypted data that are communicated and

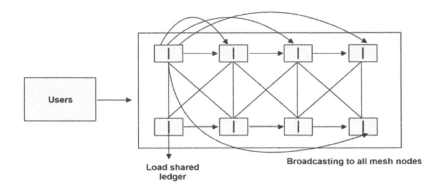

FIGURE 6.8   Registering and broadcasting mesh nodes.

broadcast accordingly to all mesh nodes. A Blockchain network consists of at least 10,000 to 20,000 nodes, and transactions of 10,000 nodes require a lot of infrastructure and power consumption. Therefore, all the transactions in a block are arranged in an array containing encrypted data with full security and transparency.

Figure 6.9 presents financial transactions with Blockchain technology with an array of blocks in a series of rows of a shared ledger.

In real time, the cryptographic puzzles will be more complex, and the solutions are sent to a network server. There is no relationship between puzzle and transaction, but the connection is commonsensical. The people who send solutions at the earliest are rewarded.

6.4.2 Data and Security Hash Algorithm (SHA 256 Algorithm)

$$H_2 = SHA256\left(H_1 + N_2 + D_2\right)$$
$$H_3 = SHA256\left(H_2 + N_3 + D_3\right)$$
$$H_4 = SHA256\left(H_3 + N_4 + D_4\right)$$
$$\vdots$$
$$\mathbf{H_n = SHA256\left(H_{n-1} + N_n + D_n\right)}$$

H_n represents the general equation with security and hash algorithm and H_1, H_2, H_3... Hn are the respective hash value, N_1, N_2, N_3.... Nn are the respective nonce, and D_1, D_2, D_3... Dn are respective data with reference

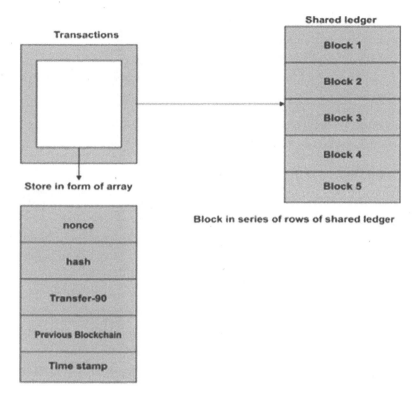

FIGURE 6.9 Pictorial representation of financial transactions with Blockchain technology.

to transactions. The above equation of hash including nonce and data is with the application of the SHA 256 algorithm. Blockchain is a tightly coupled architecture. If any data changes—for example, if D_4 changes—then H_4 changes and the whole Blockchain with it. Consensus identifies any corrupted chain, picks a correct chain, and facilitates the replacement. If the hash dependencies and hash are interconnected, which gives a chain structure, and if any changes occur, it should be identified in a fraction of the time. Proof of work in a generic form is known as a consensus algorithm. The process of performing proof of work is mining, and this process is known as cryptocurrency. If demand increases, the value increases. The job of miners is to create proof of work. The higher the amount of infrastructure generated through such confirmation, the higher the amount of bitcoin (which is a reward for creating proof of work) generated by a particular miner. Hash cash is an algorithm in which mathematical puzzles are generated. The complexity of puzzles is proportional to the size of the network. Proof of work is generally used to generate a block. There is lot of power consumption involved in generating proof of work.

1. Proof of work adds to securing the network against denial-of-service attacks.

2. Mining possibilities are plentiful, since new proof of work is constantly required.

3. High skill set is needed to solve complex puzzles in a shorter time.

4. The disadvantage is a huge expenditure in software infrastructure and power consumption.
   Proper regulation of Bitcoin currency and transactions based on it should be regularized. Cryptocurrency is a volatile mechanism prone to "bubbles." The growth of Blockchain architecture is very transient and drastic. The community wants privacy, which depends on market conditions and is generally a controversial subject. As the volume of data is increasing exponentially, the system should be fully transparent and secure. For example, we create a node and then local host 3000 completes the transactions, which establishes network nodes. It is a decentralized network containing a cluster of nodes—maybe 10,000 to 50,000—in the network.

   First client sends a request, and a node with local server 3001 is created. After the node, local server 3001 functionality is created, and then all the information is broadcast to all other nodes, from one local server to the next. Thus, every node on the network receives the client's request. Sending information from the client to all other nodes is not feasible because there are thousands of nodes, which will overload the client device and hang the app and/or browser. We will make each node capable of receiving information to other nodes in the network. Whenever web technologies are designed, the processing must be kept on the server side, not the client side. All the financial resources should be continuous and there should be zero outage. Every node should receive the transactions and should broadcast to other nodes. Therefore, for any Blockchain transactions it is essential to:

1. Establish node connection

2. Broadcast transactions to all other nodes

The first node that receives a transaction is known as a register node. For any register node there are post calls for web technologies in Blockchain. Every node should have the ability to receive information. Blockchain constructor registers the nodes in Blockchain.

5. Smart Contract. The following are the key points for the development of smart contract:

1. Identify agreement

2. Set conditions

3. Code the business logic

4. Encryption

5. Execution and updates

6. Network updates

Suppose there is a smart contract 5000 that is to be executed.
Smart Contract 5000

$$\left\{ \begin{array}{l} \text{if} \\ \text{Tr} \\ \text{else} \end{array} \right\}$$

Now, if we want to change the information in smart contract 5000, there is no way to modify the information. Blockchain is used for implementing a smart contract. Henceforth a new block will be created with the same ID of the previous block.

Figure 6.10 provides an overview methodology of smart contract for property revenue record, including all the specific features of property with insurance and internal revenue source.

Drawback of a traditional contract system:

1. Highly time consuming and expensive

2. Data leakage and limited control

3. High risk of a single point of failure and becoming a target for hackers.

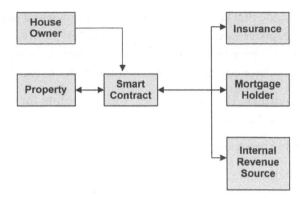

FIGURE 6.10   Block diagram of smart contract for property revenue record.

Smart contract benefits:

1. End-to-end workflow with digitization due to securities on a distributed ledger

2. Blockchain with artificial intelligence–assisted IoT for high degree of security. Blockchain has the capability for securing identities.

3. Repository for policyholder includes driving record policy, vehicle type, and accident record.

4. Vehicle self-awareness and damage assessment using sensors can execute initial insurance claims/policy reports.

5. Increased saving by reducing duplicate work to verify reports and policies.

6. Smart contract is implemented with proof of work and consensus algorithm.

Figure 6.11 shows a representation of financial transactions, smart contracts, and investor that certifies financial saving transparency and minimizes time for contract allotment procurement, processing, execution, and implementation.

6.5 CASE STUDY OF DECENTRALIZED SYSTEM

Figure 6.12 presents a case study of a decentralized Blockchain system and machine learning model in the form of an online entertainment content distributor. A global channel movie partner buys the right, the contents

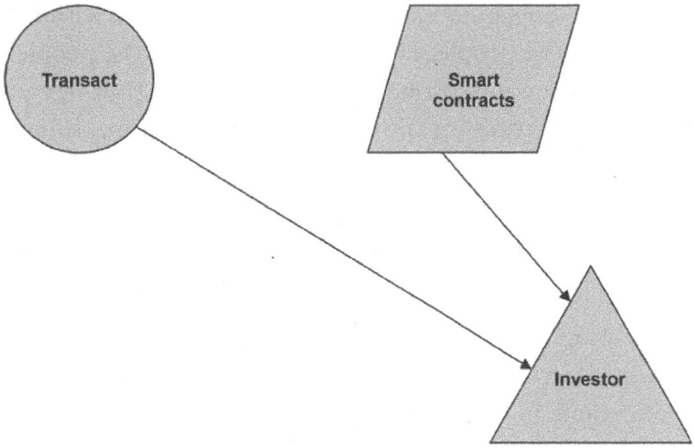

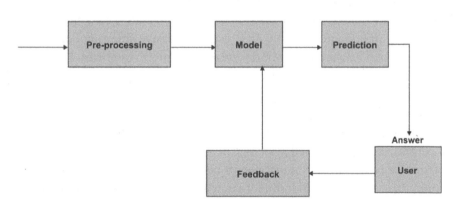

FIGURE 6.11   Block diagram of Blockchain with smart contract.

Decentralised Netflix

M-L Model

FIGURE 6.12   Block diagram of decentralised Netflix-like machine learning (ML) model with prediction analysis features.

stream on the channel, and the data (image, video, and data) are the rights of the movie channel. Users buy the subscription, which is how the movie channel partner generates revenue. In Blockchain there is no third-party intermediary and no single controller; it is a peer-to-peer model sharing information throughout the entire network. The movie channel data are uploaded into the blockchain network. There is a content delivery platform that can be developed for the decentralized network. Movie recommendation is engaging more people in the network without disclosing the identity of the people. Hence, the identity of the people will not be displayed like in a conventional system even when they like particular content. All data structures of movies are handled via machine learning, and prediction will be done in a decentralized network.

6.6 BLOCKCHAIN INFRASTRUCTURE MANAGEMENT

In Blockchain there is no cloud like in AWS or Microsoft Azure system. Content generation application is a very hot topic and an emerging area for decentralized network. Blockchain technology is a viable solution for various applications in the financial domain. Due to the complexity and costly environment to maintain, the developers face a lot of difficulties to maintain and monitor the Blockchain networks. These are different techniques that provide different solution for network deployment and system monitoring with smart contract analysis and testing. Based on the business logic, proof of work and consensus algorithm developer can focus on a distinct business scenario. A decentralized network is the one in which all Blockchain data are stored in every node and synchronized and maintained in node faults. As a result, a massive amount of data stock exchange can be shared within decentralized internet entities. Lab-based learning and modeling for stock exchange and financial transactions are done based on conventional IoT solution for neuro-informatics. The neuro-architecture informatics with all functionalities is implemented on a browser server application or client server architectures applied to a centralized server-based system. The risks of privacy breach or denial-of-service (DoS) attacks must always be considered, so in order to overcome this, a Blockchain architecture with signal processing and control is applied on a trust-free and transparent neuro-informatics mechanism. Stock exchange data are verified and stored in the Blockchain database. After verification of transactions, they are stored in the transaction pool. The constitutional link Blockchain-based internet service should be avoided, and digital signal processing will be applied to transfer the informatics system. Blockchain stock exchange data are structured in the Blockchain database after the infrastructure of transaction is stored in the transaction pool; after transaction and verification, it is coded into a new block. New signal information by an IoT device is done, and legal transaction is fetched into the block. The new block thus created is then broadcast to all selected consumer nodes with the selection of longest chain verification of each process transaction, which is done on the basis of following key points:

1. Correctness of data structure transactions.

2. Input and output should be filled and verify with definite size.

3. The value of hash as inference can be 0 or −1.

4. The communication conditions and inference should be confirmed, verified, and validated with the rules and regulation defined in the smart contract.

5. The transaction signature should be legal.

6. Confirmation of the transaction size.

6.7 DIFFERENT LAYERS OF SMART CONTRACT MANAGEMENT

The different layers of smart contract management for its execution and implementation are:

1. *Infrastructure Layer:* Infrastructure layer supports all applications of smart contract and encapsulates all the infrastructures including trusted data feeds (oracle) and trusted execution environment.

2. *Contract Layer:* It provides complete information for static contract data of all types including contract terms and conditions with its norms, rules, and regulations. It includes rules about contract invocation, execution, communication, and implementation. All meetings and discussion with contractors determine the terms and conditions of contract, which involve legal policy, business logics, and agreements. Then software programmers implement, translate, and execute the contract agreement into the program code, which is described in natural language.

3. *Operation Layer:* It includes overall operations, which are dynamic in nature on the static contracts, including verification and security analysis. The operation layer is the key layer for the correctness of the smart contract; any weak point in the contract must be verified, otherwise economic losses to customers may result. Smart contracts

at present are not so intelligent, but future smart contracts will allow for more intelligent decision-making.

4. *Intelligence Layer:* This layer includes all algorithms related to artificial intelligence with its parameters like reasoning, learning, perception, testing, training, decision-making, and socializing. It adds intelligence to the first three layers of smart contract with the development of artificial intelligence technology agents by implementation of reinforcement learning and cognitive computing.

5. *Manifestation Layer:* It includes documents of the smart contracts for potential inclusion of decentralized autonomous organization (DAO), decentralized autonomous cooperation (DAC), and decentralized autonomous society (DAS). Smart contract includes the complexities of network nodes that are interfaces of lockchain for different application scenarios. Decentralized apps (D Apps) can be developed by writing business logic and intention agreements into smart contracts.

6.8 SMART CONTRACT AND FINANCE

Smart contract increases the visibility and trust across the users, which results in considerable savings in infrastructures, transactions, and administrative cost. Different applications of smart contract in finance are:

1. *Securities:* By application of smart contract there is automatic payment of dividends and stock and liability management while reducing operation risks. By applying smart contract we can ease the clearing and settlement of securities. Blockchain encapsulates by means of smart contract and its application in business logic and its bilateral peer-to-peer application.

2. *Insurances:* Smart contract includes automatic claim settlement and processing verification and payment to increase speed of the claim processing and to eliminate fraudulent claims.

3. *Trade Finance:* At present, trade finance is highly inefficient and the industry is vulnerable to fraud. Smart contract allows businesses to automatically trigger business actions based on predefined procedures and standards that will increase efficiency by streamlining the processes and compliance and reducing fraud.

6.9 CONCLUSION

Smart contracts with Blockchain technology certify the chain of securities custody and provide ease of service for payment of dividends, stock splits, and reduced operational risk. It helps in clearing and settlement of securities. There is a concept of a three-day settlement cycle in the United States, Canada, and Japan, which involves securities depositories and management agencies. Blockchain provides mutual peer-to-peer execution and implementation of business logic using smart contracts. The insurance industry spends millions of dollars on claim processing. Smart contracts can be implemented to automate and streamline claim processing. Smart contracts can also be used in digital rights management. Blockchain technology can be used to build distributed energy system for the decentralized energy trading markets and improving energy utilization and reducing grid operating costs. Blockchain technology with smart contract can be applied in health care, production markets, and intelligent transportation systems. Smart contract is known as a research topic in industries and academia. Smart contract will change traditional financial management system by the application of IoT and artificial intelligence.

REFERENCES

[1] X. Chen, K. Zhang, X. Liang, W. Qiu, Z. Zhang and D. Tu, 2020. HyperBSA: A High-Performance Consortium Blockchain Storage Architecture for Massive Data. *IEEE Access* 8:178402–178413.

[2] W. Viriyasitavat, L. Da Xu, Z. Bi and A. Sapsomboon. 2019. New Blockchain-Based Architecture for Service Interoperations in Internet of Things. *IEEE Transactions on Computational Social Systems* 6:739–748.

[3] T. Hardjono, A. Lipton and A. Pentland. 2020. Toward an Interoperability Architecture for Blockchain Autonomous Systems. *IEEE Transactions on Engineering Management* 67:1298–1309.

[4] A. Yazdinejad, R. M. Parizi, A. Dehghantanha, Q. Zhang and K.-K. R. Choo. 2020. An Energy-Efficient SDN Controller Architecture for IoT Networks with Blockchain-Based Security. *IEEE Transactions on Services Computing* 13:625–638.

[5] S. Wang, L. Ouyang, Y. Yuan, X. Ni, X. Han and F. Wang. 2019. Blockchain-Enabled Smart Contracts: Architecture, Applications, and Future Trends. *IEEE Transactions on Systems, Man, and Cybernetics: Systems* 49:2266–2277.

[6] P. Koshy, S. Babu and B. S. Manoj. 2020. Sliding Window Blockchain Architecture for Internet of Things. *IEEE Internet of Things Journal* 7: 3338–3348.

[7] J. Wan, J. Li, M. Imran and D. Li. 2019. A Blockchain-Based Solution for Enhancing Security and Privacy in Smart Factory. *IEEE Transactions on Industrial Informatics* 15:3652–3660.

[8] T. Hardjono, A. Lipton and A. Pentland. 2020. Toward an Interoperability Architecture for Blockchain Autonomous Systems. *IEEE Transactions on Engineering Management* 67:1298–1309.

[9] M. Alharby and A. V. Moorsel. 2017. Blockchain-Based Smart Contracts: A Systematic Mapping Study. *Proceeding International Conference on Artificial Intelligence, Soft Computing*, pp. 125–140.

[10] K. Delmolino et al. 2016. Step by Step Towards Creating a Safe Smart Contract: Lessons and Insights from a Cryptocurrency Lab. *Proceeding International Conference Finance Cryptography Data Security*, pp. 79–94.

[11] D. Siegel. *Understanding the DAO Attack*. Accessed: Sep. 19, 2018. [Online]. Available: https://www.coindesk.com/understanding-dao-hackjournalists/

[12] E. Sariboz, K. Kolachala, G. Panwar, R. Vishwanathan and S. Misra. 2021. Off-Chain Execution and Verification of Computationally Intensive Smart Contracts. *2021 IEEE International Conference on Blockchain and Cryptocurrency (ICBC)*, pp. 1–3, doi: 10.1109/ICBC51069.2021.9461142.

[13] N. Atzei, M. Bartoletti and T. Cimoli. 2017. A Survey of Attacks on Ethereum Smart Contracts. *Principles of Security and Trust*. Heidelberg, Germany: Springer, pp. 164–186.

[14] R. Qin, Y. Yuan, and F.-Y. Wang. 2018. Research on the Selection Strategies of Blockchain Mining Pools. *IEEE Transactions Computer Society System* 5:748–757.

[15] T. Dickerson, P. Gazzillo, M. Herlihy, and E. Koskinen. 2017. Adding Concurrency to Smart Contracts. *Proceeding ACM Symposium Principles Distributed Computing*, pp. 303–312.

[16] G. Greenspan. *Why Many Smart Contract Use Cases Are Simply Impossible*. Accessed: Sep. 30, 2018. [Online]. Available: https://www.coindesk.com/three-smart-contractmisconceptions/

[17] F. Zhang, E. Cecchetti, K. Croman, A. Juels, and E. Shi. 2016. Town Crier: An Authenticated Data Feed for Smart Contracts. *Proceeding ACM SIGSAC Conference Computing Communication Security (CCS)*, Vienna, Austria, pp. 270–282.

[18] A. Juels, A. Kosba, and E. Shi. 2016. The Ring of Gyges: Investigating the Future of Criminal Smart Contracts. *Proceeding ACM SIGSAC Conference Computing Communication Security (CCS)*, Vienna, Austria, pp. 283–295.

[19] U.S. Securities and Exchange Commission. *Investor Bulletin: Initial Coin Offerings*. Accessed: Nov. 3, 2018. [Online]. Available: https://www.sec.gov/oiea/investor-alerts-and-bulletins/ib_coinofferings

CHAPTER 7

Revolutionizing Legal Services with Blockchain and Artificial Intelligence

Krishna Kumar Vaithinathan

Karaikal Polytechnic College, Varichikudy, Karaikal, Puducherry, India

Julian Benadit Pernabas

School of Engineering and Technology, CHRIST (Deemed to be University), Kengeri Campus, Kanmanike, Bangalore, Karnataka, India

CONTENTS

DOI: 10.1201/9781003188247-7

7.1 INTRODUCTION

Blockchain can be simply described as a decentralized method by which any type of data (including but not limited to financial transactions, securities, and asset orders) can be permanently recorded in an encrypted and irreversible ledger [1]. The first major use of blockchain technology came from the development of Bitcoin. This is a digital encrypted currency, launched in 2009 [2]. Various other digital and virtual cryptocurrencies were successively created using similar technologies. After people have widely understood that Bitcoin is based on immutable and secure technology, the use of Blockchain technology in areas other than digital payments was discussed for the first time in 2012, indicating the subsequent emergence of projects such as Ethereum [3]. Based on these discussions, potential stakeholders have begun to imagine, consider, and develop potential applications of Blockchain technology in areas other than cryptocurrencies (such as health care, transportation, intellectual property, and legal industry). As mentioned earlier, the main discussions surrounding Blockchain today are focused on how to use the technology for other potential applications (i.e., beyond digital currency) from a business and technical perspectives. One of the possible application is in the legal profession. From initial intellectual property protection to registration, licensing and enforcement, litigation, chain of custody, and settlements, Blockchain technology can be used to achieve a variety of valuable purposes. The challenge faced by this new technology, namely the commitment to the automation, reliability, efficiency, and effective protection and management system of intellectual property rights, provides an important impetus for meeting these challenges and implementing the integration of Blockchain technology in these areas.

7.1.1 Future Challenges and Hurdles

Blockchain technology has been around for less than a decade, and it is still in its infancy. However, Blockchain technology faces several challenges in four distinct fields: technology, business, education, and legal and regulatory services. In this chapter we mainly focus on potential legal/regulatory issues related to this new technology. The vital use of Blockchain as a trading platform should arouse great interest for any lawyer. This is because the Blockchain networks can verify various value transactions related to digital currency or digitized assets. In addition to digital transactions, the transactions connected to the land, debt, or intellectual property rights can also be efficiently handled by Blockchain. The current main problem is the speed at which these transactions can be processed using Blockchain technology. Compared to traditional transaction platforms (like PayPal/VISA), Blockchain is now much slower. For example, Bitcoin (the most widely used Blockchain) can process only two to five transactions per second (TPS), and Ethereum can perform at 10 TPS. Alternatively, it should be noted that many other alternatives have been introduced to solve this problem via second-layer technology [4]. They are developed over the basic level blockchains (such as Bitcoin or Ethereum blockchains).

Regarding the future legal challenges, the main obstacle that should be discussed is undoubtedly the lack of proper supervision and legal framework for Blockchain. In the past few years, the development of Blockchain technology has been much faster than expected, and a large number of applications of this technology are created in the new grey areas of prevailing/inadequate regulatory laws. Similarly, the use of various cryptocurrencies to conduct illegal transactions on the Darknet market is the first/major regulatory issue. This illegal behavior forced the regulator to quickly formulate certain rules in this area to combat money-laundering activities [5]. In addition, the second wave of regulatory uncertainty occurred during the initial coin offering (ICO), which was introduced using cryptocurrency as a method of crowdfunding. Unfortunately, despite countless ICO scams in recent years, the statements of regulators (such as the US Securities and Exchange Commission) was a vital player in the stock markets. The participation of regulators (especially securities consultants) aims to clarify the situation of ICOs and protect investors from fraud, but because the underlying technology is still in the budding stage and is developing rapidly (and not yet widely known by the regulators), it is difficult and risky to enforce regulations without fully understanding the meaning of them. Not surprisingly, the lack of comprehensive regulations has brought uncertainty to the future of Blockchain technology, and the rate at which this promising technology has been adopted by multinational companies in their daily operations has slowly declined.

Another important legal issue is the lack of planning regarding the legal requirements of early Blockchain platforms. Most of these platforms focus on trading rather than their reporting requirements. Lack of foresight forces these platforms to do value transfers to come up with new solutions for reporting problems. Although the lack of a smooth and efficient reporting process may cause severe taxation problems, it should be noted that Blockchain technology has broad prospects in alleviating the current reporting problems and significantly enhancing the reporting practices. The ongoing reporting process (such as transactions) can become trackable and irreversible by employing Blockchain technology.

7.2 LEGAL PROFESSION AND THE BLOCKCHAIN PROSPECT

At the 2017 International Legal Technology Association conference, Blockchain became the new subject of discussion. Experts are expecting that it will change the whole legal profession, having a fairly larger effect than artificial intelligence does. Figure 7.1 shows various use cases

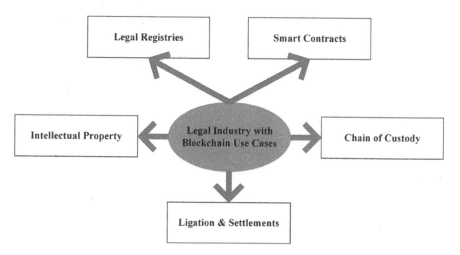

FIGURE 7.1 Legal industry Blockchain use cases.

of Blockchain in the legal profession. This exhilaration and hype normally center on the three most important traits of Blockchain:

- The removal of intermediaries for economic transactions saves more time and money.

- The tremendous distribution makes information extraordinarily secure, dependable and reliable.

- Validated information is almost indisputable.

7.2.1 Smart Contracts

The Blockchain impact of the legal industry is seen in various ways, from resolving Blockchain issues for clients to changing the distribution of legal services. Smart contracts (also called self-executing contracts) convert contracts into code that is saved, copied, and regulated by a Blockchain network. Traditionally, contract drafting requires clients to visit a law firm to discuss, negotiate, and execute agreements between respective lawyers, but smart contracts are considered to be a more time-saving and easier option. Using smart contracts, when a bitcoin is appended to the ledger, all the other required information is automatically added by the ledger. It not only defines the terms and results of the contract but also automatically executes the contract. In addition, Blockchain is a software technology that automates certain processes. This is where the popular smart contracts come into play. Smart contracts are software that resides on and cooperates with the Blockchain. The basic concept is that when a piece of particular input information fulfills the predefined conditions, certain consequences will be triggered based on algorithms (results), and these "if-then" relationships are smart contracts. Some people believe that this is non–legally binding because the parties in the contract agree that when one party performs a certain activity, the other party has to do other things. This analogy is doubtful because contracts are much more complicated. Both parties agree that they are not necessarily incomplete, and implicit exclusions can be applied in good faith. However, no matter how detailed the contract is, due diligence is not a concept that computers can use. Computer scientist Alan Turing said that computer programming languages can simulate or calculate any content word scenario that is limited only by available computing power.

7.2.2 Legal Registries

Usually the main thing that comes to mind when considering registries are those that store status, identity, or rights of someone or a legal entity, particularly industrial registers, land registries, civil registries, inheritance registers, and patent or trademark registers. Second, if blockchains are stable/secure and permit the registration of information, they may be used for securing proof that may be offered in courtroom docket proceedings. Examples are the screenshots of websites, digitized content, and course contracts. It must be noted that usually, the document itself is no longer saved at the Blockchain, but only a cryptographic hash value is stored. The capability to replicate the same hash as the one saved at the Blockchain proves that the original document has not been modified and thus is authentic.

7.2.3 Chain of Custody

The high reliability of Blockchain makes it extremely useful for building the chain of custody. With the help of Blockchain, lawyers can track the digital movement of permanently kept documents without worrying about their being altered or deleted. It can also be used to digitally tag physical objects. GPS tracking used to record every step will create irrefutable evidence for the legal cases. Blockchain affects various industries that require frequent consultation and reputation from a law firm. These innovations provide progressive law firms with new prospects to provide needed help in new areas of law. Using this, a legal professional can create a successful career on their own. According to law firms, maintaining the evidence retention chain is very difficult, but handling evidence is even more difficult. The virtual registry can be used not only to track the storage of documents but also to save documents in a very stable/safe location. Everything is documented in sequential order with a time stamp. This means that no records can be changed or deleted. No data will be lost and no testimony is required to ensure the safety of the chain.

7.2.4 Property Rights

The establishment of property rights can be very difficult, and disputes are quite common. Most records are kept in printed copies along with streams of books, journals, and similar documents, usually dating back decades. The scanned originals are archived in a database that the organization believes best suits its needs. It will take a lot of time to search in documents that are poorly archived, or records that conflict with each other.

Blockchain provides a secure, chronological, and unchanging way to store this data. This greatly simplifies the process of determining property history. All interested parties can also use the same information, thereby reducing confusion and making dispute resolution easier.

7.2.5 Intellectual Property

The digital age has brought huge intellectual property (IP) challenges, especially in terms of protecting the rights of affected people. All internet users believe that everything related to their interests should be provided for free, and the law has no solution for this issue. Illegal transfer or other use of images, audio and video files, icons, and other copyrighted materials available on the internet without permission from or compensation to their creators is rampant. Blockchain offers a way out of this conundrum by providing a publicly viewable copyright record. Anyone can track who viewed or downloaded related content. Artists can have better control over who is viewing or using their IP. Smart contracts also provide a seamless and instant system to pay for the use of materials so that creators can better protect their rights. For example, nonpayment can automatically trigger a necessary legal suit/action. The use of Blockchain can also eliminate the confusion about who owns the digital asset, making it difficult to be hacked, which also means that lawyers can quickly verify proof of ownership.

7.2.6 Financial Transactions

The unstoppable growth of cryptocurrency and online commerce supported by Blockchain technology will increasingly affect the legal profession. As more people use crypto-assets for financial transactions, they become more and more popular, and many lawyers find that they have to deal with them in day-to-day transactions. Banks, though not universally accepting or dealing with cryptocurrencies, have become more crypto-friendly and steadily enable transactions. As consumer demand increases, banks will try to adapt to crypto-assets further. Given the current dirth of regulations and legislation on the subject of crypto transactions, any lawyer has a lot of work to do to guide a bank in this rapidly changing environment. Many other novel legal services are associated with cryptocurrencies. For example, at the time of death, for fair distribution, lawyers must use crypto-assets that are stored in the Blockchain. Divorce lawyers need to find hidden assets through the system to ensure that their clients get what they are legally entitled to. Governments around the globe are looking for different ways to better integrate/regulate cryptocurrencies.

As new laws come into effect, lawyers will have to provide advice and guidance to their clients, especially about tax law.

7.2.7 Litigation and Settlements

Litigation is used to describe a legal action brought by the parties to defend or enforce the legal right. It is determined by an agreement between the parties, but sometimes by a judge or jury of the court. The litigation can range from banking to fraud, contract issues, regulatory mechanisms, and mergers and acquisitions. Litigation involves various actions that must be taken before, during, and after legal proceedings to ensure the protection of legitimate rights and interests. This can include preliminary negotiations, arbitration, legal relief, and assistance. Like other legal procedures, litigation is very time consuming and involves a lot of paperwork. Moving litigation to the Blockchain can help reduce time and increase transparency to the system. Lawyers can register themselves on the Blockchain platform by providing their details such as license number, experience, and contact information. The plaintiff can also register and hire a lawyer to provide legal aid for the dispute. After filing a claim, the defendant will receive a notice asking them to respond to the allegations. Also, the response can be sent on the Blockchain, ensuring that all information is decentralized and immutable. This makes it easier for the judiciary/jury to make a decision based on the information contained in the Blockchain. This simplifies the entire process and significantly reduces the time required to make a judgment, thereby quickly solving the case. If at any time the old case is on re-trial by a new judge, the information stored in the Blockchain can be accessed to collect all necessary documents and information again. In addition, there is no need to keep long paper records for a long time.

7.3 BLOCKCHAIN FOR JUDGMENTS EXECUTION – A PRACTICAL SCENARIO

We start our discussion by creating a psychological structure for the "phase" of Blockchain participation in the cross-border law enforcement process [6]. Generally speaking, these phases have increased the Blockchain combination levels, but these different phases overlap with each other at various levels.

- Phase 1: The traditional court judge orders a cryptocurrency compensation in his judgment, and no special combination is mentioned with the Blockchain. The parties must obtain the recognition of the judgment in the enforcement jurisdiction through the traditional recognition procedure. The law enforcement methods available in foreign jurisdictions are traditional.

- Phase 2: The traditional court makes a ruling and decides a cryptocurrency compensation. The orders/instructions are recorded in the Blockchain. Based on the judgments that are already stored in the Blockchain, judgments can be more easily identified in enforcement jurisdictions. The law enforcement methods available in foreign jurisdictions are traditional.

- Phase 3: The traditional court makes a decision and decides to pay in cryptocurrency. A smart dispute resolution contract that can automatically execute court judgments is delivered on the Blockchain, according to a Blockchain dispute resolution procedure monitored by the court. Disputes may also be recorded on the Blockchain through smart contracts. The smart contract already contains a dispute resolution agreement that prompts the court to participate, or the parties have introduced a new agreement that can automatically enforce judgments.

- Phase 4: The decision on the Blockchain is made according to the Blockchain dispute settlement procedure, which can be supervised by the court. The object of the dispute is the original transaction between the two parties to the transaction, and it may also be recorded on the Blockchain through a smart contract. The settlement agreement in the smart contract motivates the court to participate, or the parties introduce a fresh agreement that initiates/automates the execution of the judgment.

When considering each stage, this chapter discusses the programmed/spontaneous implementation of judgments (i.e., levels 3 and 4), because self-implementation of smart contracts is a distinct characteristic of the blockchain.

7.3.1 Discussion on Various Phases

Phase 1: Traditionally, the trial judge instructs the accused to pay in cryptocurrency.

Phase 2: Proof of Blockchain and authorization of overseas sentences.

Let's start with a basic scenario where court "X" judges Adam to pay Barney 100 tokens of a specific cryptocurrency. For the sake of simplicity, we assume that the court "X" is enforcing the decree and is in a jurisdiction that is friendly to cryptocurrencies.

The question arises of how to implement this judgment if the debtor fails to pay as expected. Due to the differences between jurisdictions, some traditional methods of enforcing bills of exchange include hiring debt collectors or seizing and selling property. These options can still be used for trial payments in cryptocurrency. Seizure and sale of a property can be paid in cryptocurrency/equivalents, and debts related to cryptocurrency/equivalents can be seized. If the judgment is based on the requirement for the ownership of cryptocurrency refunds, it may also force the debtor to produce the private key. If a lawsuit is filed against a cryptocurrency exchange that contains the cryptocurrency of a merchant in a wallet, it may also be forced to provide a private key if it fails to comply with a court order against its official. Traditional implementation techniques and methods are always accessible, but they have numerous disadvantages. Purchasing requests related to fulfillment are highly expensive and irritating. The above situation is compounded when a decision/judgment is made in one jurisdiction and has to be enforced in a different one. Before making an enforcement request, formal obstacles must be overcome to allow the judgment to be recognized in the new jurisdiction. Cross-border Blockchain integration may help if the proof and certification of the judgment/arbitration award can be tracked and verified in the Blockchain, and the approval stage of the judgment can be accelerated. Cross-border law enforcement The Dubai Blockchain Court Working Group is conducting a preliminary investigation into the issue. However, this is actually only a temporary improvement of administrative assistance and does not involve the essence of the problem (an effective implementation). However, if there is no applicable mutual recognition agreement (maybe for political reasons), there will be a roadblock. When the judicial system of an administrative jurisdiction is unreliable or ineffective, another level of complexity is added. As a result, in the cumbersome field of cross-border law enforcement, the automatic execution of Blockchain judgments on digital assets (such as encrypted currencies) is a new hope. Regardless of national borders, there is no need for the help of foreign courts or stamps. Therefore, it is interesting to know how the Blockchain can self-execute the cryptocurrency smart contracts.

Phase 3: Contract enforcement through smart dispute resolution

Now, we will examine the execution of judgments or decisions on the Blockchain from a practical perspective.

Situation 1: The judgment debtor (Adam in Phase 2) executes the cryptocurrency transfer transaction 100 and initiates it to Barney (judgment creditor). This is not controversial because it is the result of the debtor's initiative to take action, after the court's judgment. Situation 2: At the time of judgment, Adam and Barney have reached an agreement on the dispute resolution process and activated the immutable smart contract code of the corresponding Blockchain. The code contains instructions on how to independently transact cryptocurrencies between multiple parties if the conditions are satisfied (typically if the court decides to pay Barney from Adam's cryptocurrency wallet). In addition, due to the immutability of smart contracts, satisfaction does not depend on the resistance/postponement of the verdict pledge. If the court decision in Situation 2 is passed without using the Blockchain, then the oracles can receive the judgment data and transfer it from outside the Blockchain to the smart dispute resolution contract to initiate execution. If the decision is made by using the Blockchain to resolve disputes, then no data may be required from outside the Blockchain to activate the implementation/execution. The issues related to retrieving evaluation data from outside the Blockchain are discussed later in the chapter.

In theory, it is easy for both parties to implement smart contract code before conviction (Situation 2) to execute Blockchain judgments, but it is more important to investigate how this can be done in practice.

When the parties trigger disputes in court, they must first resolve doubts about the applicability of the legal system by analyzing the rules of conflict of laws. A Blockchain judgment cannot be made under this law. As long as this is clearly stated, they should be given access to the dispute settlement agreement implemented in the smart contract code as soon as possible. Then, compliance with this requirement is implemented and demonstrated in the early stages of the court proceedings. This operation is done in advance to ensure reliable automatic operation. This can prevent the debtor and avoid increasing the risk of indulgence before the judgment date. The parties may weigh their chances of success based on the progression of the case. If this is not done as soon as possible, the enforcement of cryptocurrency litigation will be subject to many uncertainties associated with supplementary successive litigation enforcement procedure. Basically, both parties agree to the disagreement settlement procedure, and the agreement must be registered in the court of law. This has some similarities with the Kleros Blockchain dispute resolution system. A separate preplanned transaction is used to manage the resolution process in Kleros Blockchain. In the case of a dispute over the original transaction, compensation is paid by one party to the other, but in this circumstance this hypothetical protocol is executed when the disagreement is submitted to the law court.

This is very effective for simple payment or transfer of property rights, especially when the property rights have been registered on the Blockchain. The first assessment can make the case more complicated. It is anticipated that corrective actions (a multi-step process based on prior confirmation of facts) or other temporary orders will be required. This does not necessarily exclude template code, because an advanced judicial system compatible with Blockchain can create a code repository to accomodate the maximum number of important orders/judgments. However, in the initial phases of creating this segment, new cases can profit from customized code, which can then be consistent in this segment for future reference. Filing possible legal appeals after conviction can be resolved by codifying the delay in execution related to the procedural time of appeal and taking into account the extension of the court's order. Another interesting issue is that in "normal" cases involving global components/clients, lawsuits are sometimes filed in different jurisdictions at the same time. Sometimes this requires the court to issue an order requesting the court to bring an injunction order against the party who brought the lawsuit. The parties have reached a consensus in advance on the code dispute resolution process, which links the execution of the specific process with the low probability of simultaneous trials. In a sense, this is a real-world limited jurisdiction clause. It is reasonable to say that the evaluation of the first impression, or more accurately the evaluation of the first compromise (of the parties), is the priority.

The parties can pay according to the agreement protocol, the complexity and/or measure of the issue, instead of the current administrative costs of the court. All these preliminary matters can be debated between the court and the involved parties through various meetings. This specification itself can be drafted by a professional association recognized by the government, which establishes an effective bilingual system in legal language and may be reviewed from time to time to verify its completeness, accuracy, and validity. For example, auditors have been asked to improve their skills in inspecting smart contracts and oracle. Questions are raised that if a flaw in the code causes the parties to suffer losses, they may be liable (the court and the programmer participated in the preparation of the sentence). Courts and government agencies usually enjoy certain immunity. Private service providers such as programmers are at risk, and they can purchase professional compensation insurance. However, in either case, traditional enforcement methods will also cause uncertainty, which may cause delays, such as additional courts, bailiffs/petitioners' actions, and acts of third parties. For the sake of completeness, alternative structural modifications can be made to the above discussion. In other words, if an arbitrator is approved, they can be authorized to allocate assets between various Blockchain wallets, especially under the court's authorization. Instead of relying on creating smart contract terms and obtaining institutional information from oracle to accomplish the judgment, the judge may be the one to transfer the funds. This is similar to making a payment to the court, which will be paid to the parties after the judgment is made.

If court judgments/rulings are documented on the Blockchain itself, the automatic execution of judgments will become easier, so that the details of the judgment are transferred to the Blockchain smart contract to resolve disputes on the Blockchain. This is especially important when the court decision is updated (complaint or order value). The data set contains the latest information. Of course, in cases that are too important for the courts, it is usually necessary to maintain the confidentiality of identification details or confidential information (especially when it involves arbitration awards) and the confidentiality of detailed procedural steps. The technical details of privacy protection are beyond the scope of this study. However, because the data is stored in another Blockchain or because the

Blockchain registry has not yet been accepted, it may not always be possible to deliver an in-chain dispute settlement contract with data. At least in the initial stage, it is necessary to obtain information from outside sources. When information about court decisions is stored off-chain, the relevant judiciary will sponsor an oracle to get the evidence from the secured court records. As long as there are additional intermediate stages, the risk of piracy will increase. These menaces happen both inside and outside the Blockchain world. Indeed, it may be argued that the general loss threat to Blockchain is low and can be eliminated by, for example, purchasing network insurance for Blockchain providers.

It should be noted that although the basis for court rulings may be complex, most of the orders in most final court rulings eventually involve simple payments, ownership changes, and/or various positive actions. Alternatively, the result is simple and usually easy to achieve through payment, the change of ownership (if the ownership record is stored in the Blockchain) is easy to achieve, and the order involving off-chain Blockchain assets is managed from outside the Blockchain anyway. The best solution is to deal with "mixed" judgments—for example, for products that need to meet conditions outside the chain before payment in the chain and/or products that must be done incrementally, they can still be run as part of this automatic operation. The contract is executed through a more functional smart dispute settlement agreement, and the contract will be maintained or canceled, provided that the fact that the recommended off-chain conditions are satisfied outside the network. While dealing with more off-chain conditions, there will be more potential errors and delays in the automatic execution of smart contracts. There is also the question of the physical condition confirmation. For example, if the parties must wait for the law firm to provide a paper certificate, it is safe for company officials to do so. This involves checking not only digital books outside the network but also the presence of physical objects. Perhaps all parties should prepare for the greater uncertainty inherent in mixed judgments.

Is it worth using smart contracts to automatically execute judgments?

So far, this arrangement is not much different from traditional litigation, except that most of the reasons and risks of implementation arise at an earlier stage of the process. Now let us consider a problem. In many mature Blockchains such as Ethereum, cryptocurrencies can only be stored in wallets/addresses or smart contracts agents, and then automatically issue coins to addresses after certain conditions are met, which is critical to its charm. However, if you use the example outlined above for research, it means that basically all parties in the dispute must deposit an amount that does not exceed the amount of the counterparty to ensure that the judgment is automatically executed once the judgment is passed. Claims are filed at the beginning of legal disputes, and additional reserves are provided for other cost orders, etc. This is similar to the distribution of secure benefits in court in the mode of Blockchain-based smart contracts. This restriction seems to run counter to general commercial sensitivities, especially when a large number of transactions are involved and litigation can take years (even in an effective judicial system, this is a very typical time frame). This becomes even more shocking because the value of cryptocurrencies tends to fluctuate. Usually, concerns about commercial sensitivity are discussed in the context of smart contracts, but this concern is not addressed in the context of automated law enforcement. To understand how unusual this constraint is, let's see which of them have the above limitations in traditional litigation. Perhaps these are temporary early guarantees for value and freezing orders. Compared with the number of lawsuits that need to prevent weak claims from being sued, the former is usually insignificant. They are not as important as the restrictions that are discussed, nor do they risk paying to other party. A large number of assets may become the question of the order, and the party interconnected to the order may not be able to process the frozen assets, especially if financial institutions holding these assets are notified. However, the freezing order is only approved when the risk of loss is high. In this regard, the parties are required to ensure that the execution of the freezing order is automatically executed from the start of the trial.

In a "normal" lawsuit, even if you cannot see the transaction of the counterparty's funds, you would not expect to have such an initial obligation (unlike the Blockchain, if you know the counterpart's address, you can identify the transaction in it). The provision of these funds in advance by both parties is not comforting. Moreover, the pursuit of justice is certainly very important while working to build more confidence in automatic execution when filing a claim because the restrictions were too onerous. It is believed that, fundamentally speaking, the benefits include automation of operations (as opposed to the uncertainty of traditional cross-border enforcement) and whether the dispute resolution process exceeds the disadvantages of opportunity costs during litigation and limited resources. If this is a compromise, the value proposition can be improved for all parties by increasing profits and/or reducing costs. To make more thoughtful propositions on how to enhance this value, it will be helpful to study the whole case covering the complete disagreement resolution process, and possibly even the entire dispute resolution process that takes place in the Blockchain.

Phase 4: Implementation on Blockchain following disagreement resolution by Blockchain.

There were some efforts to go into full-on Blockchain disagreement resolution, wherein generally the unique transaction, settlement, judgment delivery, and execution occur on the Blockchain itself. Also, there are numerous non-public courts and arbitration structures that include Kleros, Aragon, Jus Protocol, etc. Essentially, those structures run with the help of Blockchain jurors who determine cases with the aid of using staking tokens or with the aid of stacking up credit. National courts also are coming into the Blockchain space. Disagreements over online contracts, copyright infringements and e-commerce are revolved in the Hangzhou Court. Interestingly, this is an efficient judicial stage, wherein contracting preselected eligible parties execute a smart contract to save the transactions. On specific disagreement situations, the case is dispatched for negotiation, then for court trial. Blockchain evidence is recorded, and the involved parties can register using the platform to complement the evidence/proofs (e.g., with time-stamped screenshots of copyright-infringing websites), which is a good way to track at the Blockchain. The judicial-Blockchain analyzes the crucial risk factors and generates a final ruling that is dispatched to different organizations, such China's credit system.

So far, not much attention has been paid to the specific structure that needs to be implemented. It is said that the Blockchain court will tentatively use the Blockchain to verify the authenticity of the cross-border execution judgment (Phase 2), but in the long run, with less human effort, it strives to obey rules/laws/regulations, so that disagreements are decided on the Blockchain itself. What we can understand is that, especially in terms of implementation issues, full Blockchain integration is under development. Blockchain technology is a very ambitious project, and end-to-end processing is easier to implement when defining specific types of simple disputes that Blockchains should handle in a specific field.

7.3.2 Summary

- Cryptocurrency courts or decisions related to records on the Blockchain and the parties making such decisions can often benefit from easier access to more enforcement methods (traditional enforcement process and automated enforcement techniques on the blockchain). So the Blockchains can be efficiently used to simplify the old-style form of cross-border enforcement (Phase 2) and develop automation functions for Phases 3 and 4. The true value of Blockchain authentication is known when the claim includes a sufficient number of network effects from various jurisdictions to accept such proofs/evidences, and are reciprocal on an analogous level. The next step will be to supplement existing international agreements/ conventions to introduce Blockchain account keeping.

- As mentioned earlier, the third stage of mixed situations is likely to occur frequently, and competitive justice agencies will need to prepare for them. More and more political parties are collaborating with Blockchain or may have to use cryptocurrency and other digital media assets, especially when providing automatic operation. This is especially true for parties in or within jurisdictions where the judicial system is underdeveloped or unpredictable. Methods to completely resolve Blockchain disputes are limited and not yet fully developed, and there may not be enough bandwidth to support the argument in complex situations. It is unclear how the settlement of Blockchain disputes will have an adverse affect on the previous case values. In addition, the disputed transactions of the parties may not be completely related to the Blockchain, which requires a "hybrid" judgment in some cases.

- As time goes by, it is expected that the territorial and judicial methods of enforcement will fade. In key stages such as enforcement, due to legality reasons, arbitration's historical reliance on legal disputes will be reduced. Over time, the specialized Blockchain arbitrage in the industry will rise. In addition, the parties can choose the law and jurisdiction based on their method of resolving Blockchain disputes, which will have a more level playing field by facilitating automatic execution. Dealing with law enforcement issues is critical to choose a court.

- In creating novel technologies for operational provision systems for Blockchain disagreements and judgment enforcement, the United Arab Emirates (UAE) and China are at the forefront. There may be room for more flexible and/or younger courts to learn, adapt, and follow up on the progress made by UAE/China. With this in mind, it is noted that the advantages of implementing automated and Blockchain-compatible disagreement resolution processes overshadow the shortcomings of limited usage of reserves. Some proposals for the improvement of the value proposition of all involved parties in the Phase 3 hybrid circumstances are:

  - Speed up the mediation process. This procedure is the same as the judicial procedure in order to ensure the best interests of the parties when the case is resolved early. Every settlement is considered as a ruling/verdict and transferred to a smart contract for execution.

  - Recognizing precise areas and different types of disagreements where automated Blockchain applications may be implemented effectively. The basic laws that govern the Blockchain should be devised. The development of judicial and support structures (electronic codes) can save time and money in the long run.

  - Expedited arbitration (for example, by major international arbitration centers) can be provided for simple/smaller cases. If the amount of the claim is not properly assessed or not assessed in good faith, a procedural warning about possible major sanctions should be issued at an early stage as a soft test for the number of funds that may be restricted. An automatic cryptocurrency pegging system can be insured before approval to reduce excessive fluctuations in value.

- The parties who have gone through complicated disputes and dispute settlement procedures go to court and usually think that this will be the end, which is completely understandable. With Blockchain and well-thought-out procedures, it can be.

7.4 APPLICATION OF ARTIFICIAL INTELLIGENCE IN LAW

Lawyers are already using artificial intelligence (AI) [7], such as reviewing legal documents during litigation and due diligence, reviewing contracts that meet specified standards, conducting legal research, and predicting the outcome of cases.

7.4.1 Review of Documents

Reviewing the files to be litigated involves the task of finding related files, for example, documents with certain keywords or e-mails on a specific subject. The performance of the legal document analysis is greatly improved by AI. For example, if a lawyer who uses software with AI technology to review documents marks certain documents as relevant, AI will know which documents are relevant and will be trained accordingly. This allows AI to more accurately identify other related documents. This technique is referred to as predictive coding. Predictive coding has many advantages over traditional document analysis. They are:

- Reduce the number of irrelevant documents that lawyers must review.

- Produce statistically verifiable results that are at least more accurate than manual verification.

- Much faster than manual verification.

7.4.2 Contract Analysis

Customers need to analyze contracts in batches and individually. For instance, analyzing all the contracts that the corporation has hired can reveal risks, irregularities, future fiscal responsibilities, renewal and expiration dates, etc. This can be a slow, costly, time-consuming, and error-prone process (considering that contracts are not signed in a robust contract management system). The lawyer checks the contract every day, makes comments and revisions, and advises the client to sign the contract as it is or try to negotiate better terms. These contracts can be simple (such as NDA) or complicated. The number of contracts to study may create a bottleneck and delay transactions (and related revenue). Inexperienced legal professionals may miss some key matters, which may trouble their clients again in the future. Artificial intelligence can help review contracts, regardless of extent and complexity.

Since June 2017, JPMorgan Chase [8] has been using the AI-driven COIN process to interpret commercial credit agreements. This work when done manually took 360,000 hours; using AI it can be accomplished in few seconds. The bank also plans to use the technology for other types of legal documents. AI platforms such as Kira Systems [9] enable legal professionals to recognize, excerpt, and scrutinize commercial facts found in huge volumes of contract data. Contract summary sheets are created from those business facts to proceed with mergers and acquisitions. LawGeex [10] uses AI to analyze contracts one by one in the daily workflow of lawyers. For example, a provision in California law may be acceptable but the one in Genoa law may not. Then, when someone uploads the contract, AI will scan and determine the existing and nonexistent items. The corresponding language is highlighted according to the preset customer standards and marked with a green thumb up or a red thumb down. Company's lawyers use LawGeex to classify the standard agreements, such as a nondisclosure agreement. Contracts that meet the specified conditions can be preapproved and signed. Documents that do not meet those conditions are sent to the legal department for further review.

7.5 LEGALTECH – NATURAL LANGUAGE PROCESSING FOR LEGAL TEXT ANALYTICS

The legal linguistic should be understood from the broadest perspective: written and spoken language, legal texts, and regulatory texts such as decrees and regulations. Since linguistic and law are inextricably linked, novel natural language processing (NLP) legal methods are required to understand linguistic and legal dissertation, to develop tools that can use legal resources for legal applications and transparency via the internet to support interoperable legal systems. LegalTech [11–13] generally refers to computer technology used in various fields of legal practice and resources. An extensive range of applications exist to assist law companies and organizations in carrying out daily activities related to document support, and all procedures (legal research and document automation/compilation) and, in general, all operational aspects are associated with the dematerialization of text-based legal services to digital forms. Some of these areas are texts/documents, linguistic oriented, and directly related to legal NLP.

Therefore, language processing is essential for solving the problems and achieving goals in LegalTech. Since these tasks and goals have been solved and commercialization has been achieved, it can be said that the application of existing technologies is within reach (clearly defined text problem).

There are excellent legal information service providers, such as Thomson Reuters/Oracle, law organisations (such as Pinset Masons/Riverview Law), and many start-ups covering a wide range of topics. However, the goal of the NLP research community is to use this ability to recognize and solve complex text problems. Some of them are listed below.

NLP technology was first used to assist in the creation of legal documents. In the decision tree modeling, the document template (contract) is automatically improved and generated based on the author's local decision. This method uses fairly simple NLP technology to deliver the contract. NLP is used in large-scale legal document analysis, for example, the identification of contractual relationships that can be accessed through a collection of documents, such as global oil and gas concessions. Conventionally, law colleges, law firms, and lawmaking committees have formulated strategies that describe the interior structure and components of legal documents and how precise legal terms should be used to clarify rules and formulate decisions. In this case, NLP assists writing by governing the construction of the document, the extent of the sentence, and the usage of suggested terms. A more complex issue involves the regulation of legality/validity and uniformity of legal resources. Legal documents can be very long, are regularly updated, and are part of a huge legal system that needs to be explained and further developed based on various social and political factors. Legal entities must ensure reliability and use of latest terminology, monitor the compatibility and consistent development of norms from different jurisdictions, and verify the legitimacy of information sources and decisions/judgments related to genuine norms. In addition to a rough analysis, these comments also include a profound understanding of legal texts and reasoning. LegalDocML and LegalRuleML are one of the standards for the structure/semantic rules of legal documents.

Given the many laws and decisions passed by modern societies over the years, including economic, social, and political issues at the local and global levels, obtaining information is also a major challenge. To rebuild a house, it is good to have knowledge about the regulations in the respective area (employers should understand appropriate labor laws and the trade that is administered by global treaties/agreements). It should be possible to extract relevant texts and laws and regulations for specific cases. Most countries/regions have official websites dealing with information on laws and regulations. However, advanced search and semantic technologies are required to ensure the release, interoperability, and accessibility of these resources. There is also a need for more detailed metadata (electronic date, jurisdiction, legal issues, keywords, etc.) and detailed search/navigation software.

Retrieving and linking information is used to aid decision-making, logical reasoning, and compliance. According to laws and regulations, lawyers should be able to determine whether a particular act is legal or safe, what is the legal definition derived from the information provided, or what their responsibilities are. For example, multimedia cannot be handled/used without acquiring the various permissions attached to the project, and these permissions are usually encoded in different contracts. To ensure legal certainty, it is necessary to use tools to analyze legal documents, extract and formalize rules related to each content type, prove the rationality of the rules applicable to specific business processes, and confirm that the operation meets legal requirements and security protocols.

In addition to arguing about a specific set of legal documents, the interpretation or application of the law may also cause controversy. Arguments are usually based on legal considerations and judgments: someone wants to verify the arguments and propositions presented, and propose new arguments for and against to support reasoning. NLP technology needs to extract, collect, and link arguments from legitimate sources. The ultimate goal is to help lawyers develop, understand, and control arguments and support legal thinking.

Tools are also needed to provision related law, which is essential for cooperating with the law on a global scale. Here, it is important to fully understand the text, excerpt/cut and model the rules, and then consider these rules to check their consistency and identify errors. In addition, these transactions must be widely applicable to different legal systems, and their terms and concepts should be interrelated or consistent in all jurisdictions. Formal sources available to legal professionals and citizens and governments pose a huge challenge to NLP in terms of document development, information retrieval, knowledge representation and verification, and overall decision support.

7.5.1 Areas of Challenges for NLP

NLP itself is a developed field of artificial intelligence. It has many well-known rules-based or machine learning methods, as well as methods for separating sentences, symbolizing text, basic words, specifying words using part of the language, analyzing sentences, and enriching content. However, the content applicable to news or narrative texts does not apply to the corpus of legal language. Most legal resources are in text form:

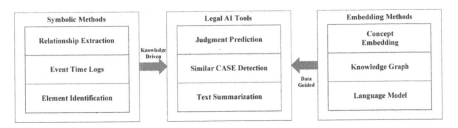

FIGURE 7.2   Overview of legal AI.

court documents, contracts, and legal opinions. Hence, most of legal AI's tasks are based on NLP technology. Many tasks in the legal field require the experience of a lawyer and a deep insight of several legal documents. Figure 7.2 illustrates the summary of legal AI. The symbolic method tends to focus on the use of legal knowledge of interpretation to distinguish symbols in legal documents, relationships, and events. At the same time, embedding-based methods try to check hidden predictive features in a large volume of data.

The dissimilarities of the two techniques indicate the drawbacks in the current process of legal AI. The explained symbolic model is not efficient, and usually the most effective embedding methods are not explainable, which may lead to moral problems in the legal system. Therefore, there is some difficulty in applying the present methods to actual legal systems due to the aforementioned drawbacks.

The three main challenges of symbol-based and embedding-based methods are:

1. Modeling Knowledge Features: The legal text has been formalized, covering many domain facts and legal AI concepts. The way of the usage of this legal knowledge is very important.

2. Legal Basis: In NLP, various procedures require proof, but the goals of legal AI are slightly dissimilar because legal proof should comply with well-defined legal rules. Therefore, the combination of predefined rules and AI technology is essential for legal defense. Complex laws and case scenarios may require more complex arguments for analysis.

3. Interpretability: Decisions prepared by legal AI must be interpretable before they are used on an actual legal system. Otherwise, justice may be compromised. This is the most important property.

7.5.1.1 Symbolic Matching/Structured Prediction Methods

The symbolic method involves the use of symbols and knowledge in the domain of legal AI. The explainability property can be achieved through symbolic legal knowledge using the events and relationships. Deep learning techniques based on symbolic methods can be used to improve performance.

- Extracting Information:

  Extracting Information (IE) has been extensively researched in NLP. IE underlines the extraction of valuable information from text. Here, two examples are given for using extracted characters to achieve legal AI interpretability:

  i. Extracting relationship and inheritance litigation:

     Inheritance disagreement is a kind of civil law case, which mainly concentrates on the distribution of the rights of inheritance. Hence, it is important to define the relationship between the parties, because the person closest to the deceased can gain more wealth. In order to achieve this goal, extracting relationships from legacy disputes can justify the outcome of the litigation and improve performance.

  ii. Extracting timeline and sentence prediction in criminal cases:

      Many parties to collective crimes often participate in group criminal cases. To determine who is predominantly accountable for the offense, it is necessary to identify the actions of everyone during the situation, and the order of events as they have occurred is also important. For example, in a crowd fight dispute, the person who initiates the fighting must take main responsibility. Therefore, the prediction of criminal proceedings judgment requires a qualified extraction model for the timeline of events.

- Extracting legal elements:

  In addition to the regular symbols in regular NLP, legal AI also has its unique symbols, called legal elements/components. Legal item retrieval concentrates on retrieving important items, such as when someone is attacked or something is stolen. Offenders can be brought to justice based on these elements. The usage of these components can not only bring temporary monitoring evidence into the evaluation and prediction problem but also make the system more interpretable.

7.5.1.2 Methods Based on Embedding Process

The embedded method emphasizes the presentation of facts and legal knowledge in the embedded latent space, and deep learning techniques can be used for associated tasks.

- Words and Concepts Embeddings:

  Embedding symbols and words play an important NLP role because they can implant distinct text in a continuous vector space. In legal AI, embedding technologies are also important because they can give more relations to texts and vectors. Also, it is impossible to understand the actual significance of conventional legal descriptions. The difficulty in learning the executive vocabulary characterization can be overcome by trying to grasp literature details and legal facts by embedding words for related tasks. Knowledge modeling is essential to legal AI because many results must be determined based on laws, regulations, and knowledge. The two main challenges of knowledge graphs in the legal domain are:

  i. Knowledge graph are important in legal AI. Mostly, the developers have to create graphs from scratch. In addition, dissimilar legal notions have diverse expressions and significances in the legal structures of various countries, which also makes it difficult for people to develop a general legal knowledge graph.

  ii. The form of the generalized table of legal knowledge is different from the commonly used form in NLP. Legal AI mainly concentrates on legal concepts explainability.

These two major challenges of knowledge modeling through embedding with legal AI are not easy to solve, and many researchers may seek ways to overcome these problems in the future.

7.5.2 Legal AI Applications

Legal AI has several typical uses, including predicting court judgments, mapping similar cases, and legal question answering. The vital functions of the civil law/common law system are predicting court judgments and mapping similar cases. Answering legal questions is the work of consultancy services for the people unacquainted with the legal domain.

- Predicting Court Decisions

  Legal judgment prediction (LJP) is one of legal AI's important tasks, especially in the civil law structure. The judgment is given based on the facts/constitutional articles in the civil law system. Legal sanctions are imposed only in the case of violations of prohibited acts as per the law. The main task of LJP is to predict the outcome of the court proceedings based on the actual explanation of the case and the content of the legal provisions in the civil justice system. Legal systems in countries such as France, Germany, Japan, and China use LJP. In addition, LJP has attracted widespread attention from AI researchers and lawyers.

- Similar Case Matching

  In the United States, Canada, India, and other common law system countries, current verdicts are based on representative cases that have been tried in the past. Therefore, identifying the most similar cases is the main challenge in the common law system. Similar case matching (SCM) has become an important topic of legal AI in predicting the outcome of common law legal disagreements. SCM concentrates on idetifying sets of analogous cases, and the description of similarities may be varied. SCM requires modeling the association between interpretations based on information of diverse granularities, such as fact level, event level, and element level. This modeling can help find the legal information easier.

- Legal Question Answering

  One of the most important parts of a lawyer's work is to deliver high-quality and reliable legal consulting services to laypersons. However, due to the inadequate number of professional lawyers, it is usually difficult to guarantee that laypersons receive sufficient-quality consulting services, and legal question answering (LQA) may solve this problem. Some questions will relate to the interpretation of certain legal concepts, while other questions may be related to the analysis of specific cases. In addition, the wording of the question can differ between professionals and non-professionals, especially when describing terms related to a particular subject area. This is one of the major challenges for LQA.

REFERENCES

[1] G. Gürkaynak, İ. Yılmaz, B. Yeşilaltay, and B. Bengi, "Intellectual Property Law and Practice in the Blockchain Realm," *Comput. Law Secur. Rev.*, vol. 34, no. 4, pp. 847–862, Aug. 2018, doi: 10.1016/j.clsr.2018.05.027.

[2] V. Gupta, "*A Brief History of Blockchain,*" 2017. https://hbr.org/2017/02/a-brief-history-of-blockchain (accessed May 04, 2021).

[3] "*Ethereum Foundation|ethereum.org,*" 2021. https://ethereum.org/en/foundation/ (accessed May 04, 2021).

[4] A. van Wirdum, "*Segregated Witness Activates on Bitcoin: This is What to Expect—Bitcoin Magazine: Bitcoin News, Articles, Charts, and Guides,*" 2017. https://bitcoinmagazine.com/technical/segregated-witness-activates-bitcoin-what-expect (accessed May 04, 2021).

[5] L. M. (UCL) Nejc Novak, "*EU Introduces Crypto Anti-Money Laundering Regulation|by Nejc Novak, LL.M. (UCL)|Medium,*" 2018. https://medium.com/@nejcnovaklaw/eu-introduces-crypto-anti-money-laundering-regulation-d6ab0ddedd3 (accessed May 04, 2021).

[6] C. Zhen Er Low, "*Execution of Judgements on the Blockchain—A Practical Legal Commentary,*" *Harv. J.L. Tech. Dig.*, 2021, [Online]. Available: https://jolt.law.harvard.edu/digest/execution-of-judgements-on-the-blockchain-a-practical-legal-commentary.

[7] L. Donahue, "*A Primer on Using Artificial Intelligence in the Legal Profession—Harvard Journal of Law & Technology,*" 2018. https://jolt.law.harvard.edu/digest/a primer-on-using-artificial-intelligence-in-the-legal-profession (accessed May 04, 2021).

[8] H. Son, "*JPMorgan Software Does in Seconds What Took Lawyers 360,000 Hours|The Independent|The Independent,*" 2017. https://www.independent.co.uk/news/business/news/jp-morgan-software-lawyers-coin-contract-intelligence-parsing-financial-deals-seconds-legal-working-hours-360000-a7603256.html (accessed May 05, 2021).

[9] "*Machine Learning Contract Search, Review and Analysis Software|Kira Systems.*" https://www.kirasystems.com/ (accessed May 05, 2021).

[10] "*Homepage—LawGeex.*" https://www.lawgeex.com/ (accessed May 05, 2021).

[11] P. Callister, "Law, Artificial Intelligence, and Natural Language Processing: A Funny Thing Happened on the Way to My Search Results," *Law Libr. J.*, vol. 112, p. 161, Oct. 2020.

[12] A. Nazarenko and A. Wyner, "Legal NLP Introduction," *Assoc. pour le Trait. Autom. des Langues*, 2017.

[13] H. Zhong, C. Xiao, C. Tu, T. Zhang, Z. Liu, and M. Sun, "*How Does NLP Benefit Legal System: A Summary of Legal Artificial Intelligence,*" arXiv, Apr. 2020, accessed May 05, 2021. [Online]. Available: http://arxiv.org/abs/2004.12158.

CHAPTER 8

An ML-Driven SDN Agent for Blockchain-Based Data Authentication for IoT Network

Rohit Kumar Das
VIT-AP University, Andhra Pradesh, India

Sudipta Roy
Assam University, Assam, India

CONTENTS

8.1 INTRODUCTION

In recent years, the Internet of Things (IoT)network has gained much attention in terms of its usability and durability. The IoT network has changed the working procedure of various day-to-day applications in today's world [1]. Being a part of a wireless sensor network (WSN), the easy installation,

DOI: 10.1201/9781003188247-8

openness, and low-cost sensor nodes have changed the perception of using WSN. The IoT network uses resource-constraint sensor devices to collect the data and transmit them to the centralized cloud server for further analysis. Though an IoT network can be implemented with limited resources, it still faces some prominent challenges such as reliability, availability, scalability, interoperability, security, and privacy [2].

To further strengthen the working of IoT networks, there has been an effort made to use the advantage of software-defined network (SDN) with IoT networks. SDN can help monitor the network with limited intervention. SDN separates the data plane and the control plane of any network entity. The controller is kept at the control layer responsible for network maneuvering, whereas the data plane has the functionality to transmit data according to the rules provided by the controller [3]. As shown in Figure 8.1, the recent development in the SDN is multiple controllers in the control plane. The data plane of the SDN network consists of OpenFlow switches that are programmable and store the flow rules. The flow rules are installed on the OpenFlow switch by the controller through the southbound interface (SBI). In case any alternation is required, a user can access the network using the northbound interface (NBI) and perform the necessary actions.

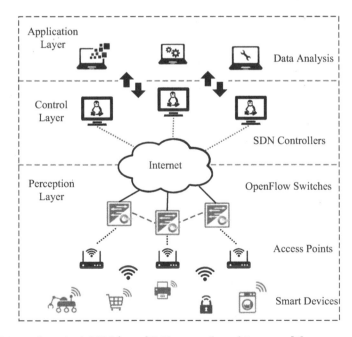

FIGURE 8.1   A typical SDN-based IoT network architecture [4].

As the IoT devices are connected over the internet, they are accessible from anywhere in the world. This gives rise to the issue of authenticity and integrity of the data being transmitted by the connected IoT devices. Because of the resource-constrained nature of the IoT devices, traditional security mechanisms are not suitable, as they require more resources [5]. Moreover, due to the prominent challenges of IoT networks such as reliability, availability, scalability, and interoperability, it becomes a challenging task to provide a global optimal solution for the security of the data.

8.1.1 Motivation

The transmission of data from the IoT gateway node to the cloud infrastructure can be vulnerable to intruders. As discussed earlier, due to the resource-constrained nature of IoT devices, it is not feasible to host security applications directly with the IoT devices [6]. The SDN controller is powerful and has the capability to host the necessary security mechanism. The SDN controller and the IoT gateway device [7] can be configured to limit the unauthorized third party's access to data.

Blockchain was introduced as a scalable security solution for the IoT network. Blockchain is a distributed ledger technology (DLT) that enables distributed computing facilities with storage, authenticity, and integrity of different transactions made in the network [8]. Furthermore, incorporating Blockchain with distributed SDN controllers can significantly increase the security of IoT networks [9].

Machine learning algorithms have proved to be an adequate measure that can enhance the security level of IoT data through proper validation of data using variously supervised and unsupervised learning approaches [10]. Based on the type of attacks, counterdefense policies can be designed for the IoT system, and the intruder can be detected at the early stage [11]. Reinforcement learning in IoT [12] is one such approach that monitors the environment continuously and generates feedback. Based on the feedback, it can make decisions of how to deal with the associated challenge.

The volume of data generated by IoT systems is inherently high, and maintenance of this volume becomes a challenging task [13]. To overcome this issue, deep learning has shown promising results, which ensures proper validation of data by incorporating testing and training models [14]. The deep learning application can be hosted at the cloud layer where the required amount of resources can be provided for further processing.

8.1.2 Contribution

In this chapter, we propose a Blockchain-based distributed SDN architecture for the IoT network. It can help overcome most of the challenges faced by the traditional IoT network. The use of Blockchain and SDN resources for IoT networks can significantly improve network performance. Moreover, to make the network self-sustainable and intelligent, a machine learning (ML) approach has been introduced. The ML-Agent will run along with the SDN controllers that will be responsible for authentication of the data. The proper validation of the data will be carried out by the Blockchain network.

The rest of the chapter is organized as follows. Section 8.2 reviews the related work in this domain. The proposed architecture with an ML-driven SDN agent is presented in Section 8.3. Potential research scope is presented in Section 8.4. Section 8.5 offers a conclusion.

8.2 RELATED WORK

There have been efforts made by various researchers and practitioners to provide efficient security measures for an IoT network.

BlockSecIoTNet [15] is an IoT architecture that provides security measures for IoT networks using Blockchain. It monitors for anomalous traffic flows by using SDN controllers that are placed in the distributed Fog network. The cloud layer is responsible for further data processing, which is transmitted to the SDN controller to take necessary actions. The architecture provides security features in the Fog layer by using a Blockchain module that instructs the SDN switches to detect specious flows in the network.

The authors of [16] offer a provision of data traffic classification in SDN networks using ML. It classifies and provides the accuracy of prediction for the flows using support vector machine (SVM), Naive Bayes, and nearest centroid approaches. The study also focuses on incorporating reinforcement learning in the SDN network where the controller acts as an agent that learns the behavior of the network from the collected flows. The proposed model uses the collected data from the various applications and performs offline classification.

BFF-IDS [17] was proposed for intrusion detection in SDN-enabled vehicular networks. Blockchain in BFF-IDS is used for storing the locations and hash values of the vehicles. The control layer consists of Blockchain miner nodes that create blocks, and verification is processed via proof-of-authority. The architecture intends to provide security using federated learning that creates a trained model for accessing the vendor specification confidentially.

BMC-SDN [18] is a multiple SDN controller network architecture based on Blockchain. The architecture ensures security measures in the control layer by creating multiple domains of the SDN network. The control layer is an interconnection of multiple SDN controllers that are hosted as a part of the Blockchain network. Each network consists of a master controller responsible for the creation of blocks, while the remaining controllers validate the respective blocks. The block validation is performed by using a threshold value by the consensus protocol.

DRQN-NAF [19] is based on a reinforcement learning mechanism that provides a distributed implementation of Software-Defined Industrial Internet of Things (SDIIoT). The Blockchain is implemented based on the Markov decision process. It uses Blockchain to enable synchronization between multiple controllers to achieve a common global view of the network and provides features for cryptographic operations. Deep reinforcement learning is used to set up the model for the Markov decision process that optimizes the Blockchain-based SDIIoT.

The authors in [20] present a distributed secure architecture SDN-based IoT for smart cities. It uses Blockchain for providing security against cyberattack. The Blockchain hosts the blocks that are connected to the single SDN controller. The network layer is formed by using gateways and SDN-enabled switches. The security mechanism is implemented for data that are forwarded from the gateway to the cloud. Moreover, to reduce energy consumption, it utilizes a cluster head selection approach. Multiple cluster heads are selected within the network domain that transfers the data to the gateway device.

It is evident from the literature that using advanced technologies such as SDN, Blockchain, and ML has enhanced the security features of existing IoT networks. In the proposed SDN-based IoT architecture, the incorporation of an ML agent in the SDN controller layer will create the ability to detect any malicious flows in the network. The Blockchain in the application layer can provide data integrity and authenticity by using the smart contract that will be responsible for traffic analysis and classification.

8.3 PROPOSED ARCHITECTURE

The proposed architecture is presented in Figure 8.2. It consists of three networks, namely the IoT network, SDN network, and Blockchain network. As shown in Figure 8.2, the access points connect the smart devices

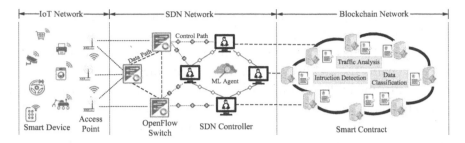

FIGURE 8.2   Proposed ML-driven SDN agent for SDN-based IoT network using Blockchain technology.

from the IoT network to the SDN network. All network operations such as routing, management, and updating policies are performed by the SDN controllers. The SDN controller configures the routing flow by using the OpenFlow protocol. It installs flow rules in the OpenFlow switch, which are used for the transmission of data.

To authenticate the flow of the network, ML-Agents are configured with the SDN controllers that are responsible for maintaining the authenticity and integrity of the flows. These SDN controllers are interconnected to each other so that the network can sustain the scalability of the IoT network. The ML-Agent continuously monitors for suspicious flow entries, and upon detection, the SDN controllers are notified. All the collected data are further transmitted to the Blockchain network for necessary analysis, creating a record pattern of malicious activity and classification. The nodes in the Blockchain network are configured with powerful resources that use smart contracts for the proper validation of the detected suspicious activity and store them for future reference.

The ML-Agents employed to monitor the data flows work on the basis of reinforcement learning principle. They continuously monitor the network and collect the feedback. The feedback is further analyzed to detect any suspicious flow that can compromise the normal flow of data. A reward is issued if any suspicious flow entry is found in the flow table and the SDN controller is notified. The SDN controller using the control path modifies or updates the flow entries in the OpenFlow switches. To maintain the consistency of the network state, the same action is notified to the other SDN controllers that are working together in synchrony with each other.

The detection of the malicious flow can be achieved by checking the property of the flow entries. The ML-Agent is moved to the next flow table

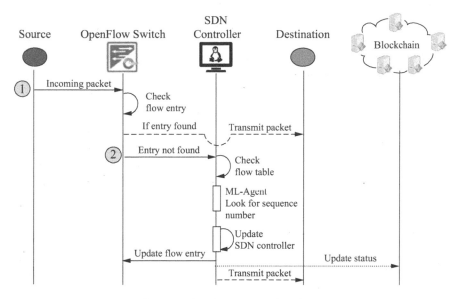

FIGURE 8.3   Sequence diagram for the proposed architecture.

if and only if the sequence of the input packet in the current flow table is greater than the previous value, and the countervalue is updated. As presented by case 1 in Figure 8.3, the packet is forwarded to the destination when it matches the flow entry in the OpenFlow switch. In case the flow entry is not found or there is a mismatch in the sequence number of the flow table as shown in case 2 in Figure 8.3, the SDN controller is notified. Based on the network condition, the SDN controller will be responsible to take the action on whether the packet must be forwarded to the destination or be discarded. The action taken on the mismatched packet will be forwarded to the Blockchain network for making future decisions.

The proposed method provides an efficient and significant approach for the detection of any suspicious flow in the network that can compromise the integrity and authenticity of data. As discussed in the previous sections, the SDN controller is a powerful resource entity that can provide all the necessary requirements for the configuration of ML-Agent. The use of Blockchain in the proposed architecture further strengthens the network by providing adequate resources for proper validation of the data.

8.4 POTENTIAL RESEARCH SCOPE

The use of ML approaches for increasing demand in security features of SDN-based IoT architecture is addressed by various researchers. Data integrity and authenticity are one of the major requirements for any open network such as IoT. Though significant improvement has been provided by various research, they lack a significant amount of challenges and open issues that require an effective solution. This gives rise to various open research directions, which are listed below.

1. The traditional SDN architecture is a centralized architecture with a single controller in the control plane. Recent advancement with distributed SDN controllers has proved that multiple controllers can increase the network performance. The use of multiple controllers will require a proper synchronization and resilience mechanism so that they can efficiently share and validate the data.

2. The consensus protocols for the SDN-based agent can be implemented in the different layers of the SDN-based IoT architecture to overcome the data integrity issue. The working principle of the consensus protocols is greatly influenced by the nature of the application in which the system will be deployed. The use of ML will also trigger the adaption of consensus protocol with IoT systems.

3. As seen from the literature, the use of a ML approach can greatly increase IoT system performance in terms of security. Due to the resource-constrained nature of the IoT devices, the use of traditional security algorithm of ML will require a very high amount of resources. The provision of developing a lightweight security mechanism for IoT networks is already getting attention from various perspectives.

4. As each layer will be handling the data, proper classification of security measures is required to secure the data at every layer of the network. In the lower perception layer and middle network layer, an on-demand security mechanism is essential, which forwards only the authenticated data to the upper layer. This will ensure lesser resource utilization and an effective solution for maintaining the integrity of the forwarded data.

5. The incorporation of Blockchain technology with IoT networks has shown its potential for providing security measures in recent years. It may be noted that the use of Blockchain in the several layers of IoT can be a difficult task, as IoT nodes are resource-constrained in nature. In this context, lightweight Blockchain mechanisms are gaining significant attention, which can be used efficiently for

maintaining the integrity and authenticity of the data in every layer of the IoT network.

6. Based on the nature of applications, learning procedures such as supervised and unsupervised learning can be adopted. The design of the security principle should be focused on the data being transmitted over the network. The current scenario appeals to more hybrid approaches that can provide sustainable security measures. The machine learning algorithms should be developed in such a way that they only learn the required amount of detail to conserve the resources of the IoT devices.

8.5 CONCLUSION

The incorporation of machine learning security mechanisms has resulted in providing enhanced secured IoT networks in recent years. This endeavor has yet to gain an industrial standard. This chapter has studied the incorporation of machine learning in SDN-based IoT networks. In this chapter, an agent-based ML architecture was proposed to provide security in terms of maintaining data integrity and authenticity for SDN-based IoT networks. The agent learns the flow rules and alerts the SDN controller in case of any suspicious flow found in the flow table. This restricts the intruder's ability to modify data in the perception and control layers of the network. Based on the studies, this chapter also provided further research directions in the area of how IoT networks can be made more sustainable and resilient in terms of maintaining the integrity and authenticity of the data.

REFERENCES

[1] Leonardo Babun, Kyle Denney, Z. Berkay Celik, Patrick McDaniel, and A. Selcuk Uluagac. A survey on IoT platforms: Communication, security, and privacy perspectives. *Computer Networks*, 192:108040, 2021.

[2] Abbas Shah Syed, Daniel Sierra-Sosa, Anup Kumar, and Adel El-Maghraby. IoT in smart cities: A survey of technologies, practices and challenges. *Smart Cities*, 4(2):429–475, 2021.

[3] Zahra Eghbali and Mina Zolfy Lighvan. A hierarchical approach for accelerating IoT data management process based on SDN principles. *Journal of Network and Computer Applications*, 181:103027, 2021.

[4] Rohit Kumar Das, Nurzaman Ahmed, Fabiola Hazel Pohrmen, Arnab Kumar Maji, and Goutam Saha. 6LE-SDN: An edge-based software-defined network for internet of things. *IEEE Internet of Things Journal*, 7(8):7725–7733, 2020.

[5] Wu Jun, Mianxiong Dong, Kaoru Ota, Jianhua Li, and Wu Yang. Application-aware consensus management for software-defined intelligent blockchain in IoT. *IEEE Network*, 34(1):69–75, 2020.

[6] Oludare Isaac Abiodun, Esther Omolara Abiodun, Moatsum Alawida, Rami S. Alkhawaldeh, and Humaira Arshad. A review on the security of the internet of things: Challenges and solutions. *Wireless Personal Communications*, 119(3): 1–35, 2021.

[7] Kübra Kalkan. SUTSEC: SD Nutilized trust based secure clustering in IoT. *Computer Networks*, 178:107328, 2020.

[8] Li Xu Da, Yang Lu, and Ling Li. Embedding blockchain technology into IoT for security: A survey. *IEEE Internet of Things Journal*, 8(13):10452–10473, 2021.

[9] Darshan Vishwasrao Medhane, Arun Kumar Sangaiah, M Shamim Hossain, Ghulam Muhammad, and Jin Wang. Blockchain-enabled distributed security framework for next-generation IoT: An edge cloud and software-defined network-integrated approach. *IEEE Internet of Things Journal*, 7(7):6143–6149, 2020.

[10] Sherali Zeadally and Michail Tsikerdekis. Securing internet of things (IoT) with machine learning. *International Journal of Communication Systems*, 33(1):e4169, 2020.

[11] Syeda Manjia Tahsien, Hadis Karimipour, and Petros Spachos. Machine learning based solutions for security of internet of things (IoT): A survey. *Journal of Network and Computer Applications*, 161:102630, 2020.

[12] Fatima Hussain, Rasheed Hussain, Syed Ali Hassan, and Ekram Hossain. Machine learning in IoT security: Current solutions and future challenges. *IEEE Communications Surveys & Tutorials*, 22(3):1686–1721, 2020.

[13] Sushil Kumar Singh, Young-Sik Jeong, and Jong Hyuk Park. A deep learning-based IoT-oriented infrastructure for secure smart city. *Sustainable Cities and Society*, 60:102252, 2020.

[14] Mohamed Ahzam Amanullah, Riyaz Ahamed Ariyaluran Habeeb, Fariza Hanum Nasaruddin, Abdullah Gani, Ejaz Ahmed, Abdul Salam Mohamed Nainar, Nazihah Md Akim, and Muhammad Imran. Deep learning and big data technologies for IoT security. *Computer Communications*, 151:495–517, 2020.

[15] Shailendra Rathore, Byung Wook Kwon, and Jong Hyuk Park. Blockseciotnet: Blockchain-based decentralized security architecture for IoT network. *Journal of Network and Computer Applications*, 143:167–177, 2019.

[16] Meenaxi M Raikar, SM Meena, Mohammed Moin Mulla, Nagashree S Shetti, and Meghana Karanandi. Data traffic classification in software defined networks (SDN) using supervised-learning. *Procedia Computer Science*, 171:2750–2759, 2020.

[17] Ibrahim Aliyu, Marco Carlo Feliciano, Šelinde Van Engelenburg, Dong Ok Kim, and Chang Gyoon Lim. A blockchain-based federated forest for SDN-enabled in-vehicle network intrusion detection system. *IEEE Access*, 9:102593–102608, 2021.

[18] Abdelouahid Derhab, Mohamed Guerroumi, Mohamed Belaoued, and Omar Cheikhrouhou. BMC-SDN: Blockchain-based multicontroller architecture for secure software-defined networks. *Wireless Communications and Mobile Computing*, 2021: 107328, 2021.

[19] Jia Luo, Qianbin Chen, F. Richard Yu, and Lun Tang. Blockchain-enabled software-defined industrial internet of things with deep reinforcement learning. *IEEE Internet of Things Journal*, 7(6):5466–5480, 2020.

[20] Md Jahidul Islam, Anichur Rahman, Sumaiya Kabir, Md Razaul Karim, Uzzal Kumar Acharjee, Mostofa Kamal Nasir, Shahab S. Band, Mehdi Sookhak, and Shaoen Wu. Blockchain-SDN based energy-aware and distributed secure architecture for IoTs in smart cities. *IEEE Internet of Things Journal*, 22(1): e4169, 2021.

CHAPTER 9

IoT-Enabled Peer-to-Peer (P2P) Trading of Rooftop Solar (RTPV) Power on Blockchain Platform in India

Shuvam Sarkar Roy

The World Bank, New Delhi, India

CONTENTS

DOI: 10.1201/9781003188247-9

9.1 INTRODUCTION

Blockchain is a modern groundbreaking technology built on a digitally distributed ledger for keeping permanent and tamper-proof (immutable) records of transactional data that has piqued the attention of energy companies, national governments, and academic institutions. When integrated with smart contracts, Blockchain offers clear, encrypt, and secure services that can allow new business solutions [1].

Why Is the Blockchain Platform Used for Peer-to-Peer (P2P) Trading?

Blockchain technology is an ideal P2P trading marketplace because it brings transparency and keeps a record of all transactions taking place on the network. All information transferred via Blockchain is encrypted, and any manipulation of the data is immediately detectable by all other parties. As transactions are recorded at every step, Blockchain ensures security, transparency, and reliability of the data.

In P2P trading, individuals generate energy from rooftop solar (RTPV) and trade the excess energy (after personal consumption) with other consumers, at a price agreed between prosumer and consumer. This tamper-proof data, which does not originate from DISCOM, is a critical market enabler.

9.2 CONTEXT

By 2022, the Ministry of New and Renewable Energy (MNRE) planned to install 40 GW of solar rooftop (RTPV) power. [2]. As of March 2021, India had about ~4.5 GW of RTPV installed capacity [3], and it is anticipated that the government will continue to promote solar rooftop program. Solar generation is expected to scale up faster due to the following factors:

a) Cost reduction trajectory of solar PV modules witnessed in the recent past is expected to continue in the coming years while cell efficiency is expected to increase.

b) Rooftop PV is financially beneficial to most categories of consumers.

c) Inclination of consumers toward green energy as awareness is increasing.

d) Consumers will be interested to achieve self-reliance in energy terms.

e) Policymakers are already incentivized and have recently noticed the adoption of gross metering for load >500 kW [4]. This has increased the interest of larger customers in exploring alternative approaches to realize savings on electricity bills from a RTPV investment, including P2P trading of energy from rooftop PV systems.

According to the current trends, the grid will become more interactive. A P2P trading solution is an essential move on the road to a more interactive grid that benefits customers, prosumers, and DISCOMs. Due to the P2P power-trading paradigm, consumers now have increased access to renewable energy. This allows them to make greater use of their distributed energy resources. In addition, an area's resistance to power outages in an emergency is improved, and in some situations, energy availability is increased.

9.3 OPPORTUNITIES FOR TRADING

The P2P trading model provides an online platform where power is traded between prosumers and consumers at a mutually specified price without the need for an intermediary using IoT-enabled smart meters. Further, P2P trading has the potential to provide following benefits for a DISCOM:

a) *The empowerment of consumers and prosumers is promoting renewable energy and flexibility:* P2P trading platforms can provide a marketplace for prosumers to exchange renewable energy produced at a lower cost, facilitating distributed generation deployment. Similarly, P2P trading gives customers more leverage over their energy use and price, increasing system flexibility. If storage facilities are implemented, participants in P2P trading can boost their local communities by allowing them to absorb renewable energy and gain more from their distributed generation. Customers without renewable generation facilities will profit directly from local renewable generation through P2P trading.

b) *Balancing and congestion management on the distribution grid through better operation of distributed energy resources:* P2P trading systems increase management of decentralized generators by matching local power demand and supply. P2P trading will considerably reduce the amount of money spent on generating capacity and transmission infrastructure to meet peak demand, resulting in lower aggregate technical and commercial (AT&C) losses due to reduced congestion on transmission and distribution networks, in addition to increasing local renewable energy usage.

c) ***Integration of Ancillary Services to the Grid:*** In addition to enabling P2P transactions, P2P network operators can allow participants to supply ancillary services to the main grid if prosumers use energy storage systems. Due to tremendous potential of scaling up electric vehicles (EV) across the country, the vehicle-to-grid (V2G) technologies will facilitate aggregation of large number of EV batteries and other distributed energy storage systems at prosumer premises as virtual power plant (VPP). VPPs can also provide grid ancillary services.

9.4 SOLUTIONS FOR P2P TRADING

The DISCOM (or a service provider) can promote local generation of solar energy by lending money to customers to create RTPV infrastructure under the operational expenditure (OPEX) or capital expenditure (CAPEX) model and creating an end-to-end, tamper-proof centralized P2P trading and settlement system for commercial settlement of locally generated energy. IoT-enabled devices such as smart meters are an imperative in enabling a seamless functioning of P2P transactions and integration with existing utility systems. In order to roll out such a system, the following things are necessary:

a) Software platform for commercial settlement – metering, billing, collection (MBC)

b) Established value chain for solar installations

c) Digitalization: along with the physical layer of P2P electricity trading (e.g., minigrids, microgrids, distribution networks, etc.), a virtual or digital layer is required for this business model. An aspect of it is an energy management system (EMS) on the platforms that enable many peers to connect and facilitate P2P trading. To ensure the power system's efficiency, data from both producers and customers must be obtained and analyzed. Smart meters, communication infrastructure, and network and automation systems (digitalization) are critical enablers of platform-based business models like P2P power trading.

d) IoT-enabled smart meter communicating with the Blockchain platform

e) Wallet/UPI-based payment solution

9.5 BENEFITS FOR STAKEHOLDERS

9.5.1 Benefits for Participants (Prosumers and Consumers)

i. Freedom for the consumer to choose a supplier of energy (green energy)

ii. Flexibility for customers to trade net energy available at a given time at a better rate (than the net metering or gross metering tariff) to maximize revenue

iii. Prosumers can monetize their excess solar energy in the same billing cycle

iv. Incentives from DISCOM for participation in demand-side management (DSM)

v. Participants can sell their energy at market-determined price

vi. As no minimum or maximum capacity for renewable generation is specified in the system, even prosumers with very small RTPV systems or very large aggregators can join the P2P trading platform.

9.5.2 Benefits for DISCOMs

i. RPO fulfillment for DISCOM: provided the regulations allow for all rooftop PV generation to be counted toward RPO, regardless of whether it is sold directly to the DISCOM or to a neighbor, or self-consumed.

ii. Increased generation from rooftop solar reduces the quantity of electricity to be transmitted to local communities, which in turn reduces the distribution losses as well as defers investment on a system upgrade.

iii. Reduction in the procurement of excess rooftop solar energy from consumers: by enabling RTPV energy to be traded directly between prosumers and consumers, DISCOMs need not buy the RTPV energy from prosumers under regulated net metering tariffs, which are usually higher than the average power purchase cost of the DISCOM.

iv. In the absence of energy storage systems, the prosumer can be encouraged to sell the excess generation of RTPV energy wherever they get the best price through P2P trading, as they would be unable to manipulate monthly billing slabs to limit monthly consumption just below higher slabs in this situation (which is one form of tariff arbitrage). Even if a customer continues to minimize monthly consumption by P2P trading to stay below higher tariff slabs in the cost-plus tariff system, such revenue losses can be balanced in the following year in the tariff petition.

v. Potential new revenue streams for the DISCOM: DISCOMs can levy wheeling charges and billing and transaction fees for the energy traded on the P2P trading platform within their distribution network.

vi. Advocating usage of Blockchain-based technologies would ensure rapid adoption of smart meters, which eases the metering, billing, and collection (MBC) process (efficiency and transparency). A Blockchain solution identifying the energy source, at what unit price and any mark-up passed to the consumer, would result in more competitive pricing.

vii. Balancing local generation and demand: DISCOMs can enable prosumers and consumers to discover and regulate their energy usage through preparation of buy and sell orders to balance the demand and supply within the local community.

viii. Voltage and capacity constraint management: promotion and consumption of RTPV energy in local communities can prevent voltage and reverse power flow issues.

ix. Reduction in DISCOM portfolio requirement results in less procurement of power, which in turn creates savings for all consumers, even those who are not part of the P2P trading scheme.

x. Enabling orderly scaling-up of renewable energy on the grid.

9.6 PROJECT OVERVIEW

Uttar Pradesh (UP) is the first Indian state to establish a pilot project to use Blockchain technology to trade electricity generated by customers' RTPV among themselves with other consumers who do not have RTPV.

Under the existing scenario, the surplus electricity generated from RTPV can be sold back to the electricity supply company (MVVNL in case of Lucknow) only at the regulated price of Rs2 per unit (kWh) in UP [5]. The successful outcome of the pilot project enabled the prosumers (customers with RTPV) to trade their surplus power produced from RTPV to other customers (peers) at a market-driven price, which is higher than the regulated rate of Rs2/kWh. This pilot project was conceived by the Honorable UP Electricity Regulatory Commission (UPERC). The Uttar Pradesh Power Corporation Limited (UPPCL) and its subsidiary Madhyanchal Vidyut Vitran Nigam Limited (MVVNL) are hosting this pilot project in Lucknow. UPERC allowed the pilot project to be implemented under the regulatory sandbox approach to test the technical feasibility and customers' willingness to participate in such programs. The trading during the pilot phase was undertaken as mock trade—no actual money transactions were involved. But the platform could demonstrate how much the project participants would have gained if there was actual money flow.

The project is implemented by ISGF in collaboration with Power Ledger, Australia. MVVNL and ISGF jointly identified and recruited 12 participants in the pilot project—9 of them with RTPV and 3 of them net buyers. ISGF installed smart meters in series with the existing meters of MVVNL. The ISGF smart meters (made by Crystal Power) were integrated with the Blockchain platform of Power Ledger. These meters have 4G SIM cards of Vodafone Idea Ltd. Abjayon Inc., a system integrator, has helped in integration of the Blockchain platform with UPPCL's billing system. ISGF and Power Ledger conducted trainings for concerned officials of UPPCL, MVVNL, and nodal officers/owners of the identified buildings on the features and functionalities of the P2P platform and the trading procedures.

The key objectives of the project were to assist all parties in understand the nuances of P2P trading such as:

i. To test the technical feasibility of P2P energy trading platform and value proposition to MVVNL and their prosumers and consumers.

ii. The potential value that can be derived from selling RTPV energy to other buildings within the area.

iii. The value of a dynamic energy marketplace where prosumers and consumers can set their buy and sell prices.

iv. Test customers' willingness to participate in P2P trading programs.

v. The optimization that can occur from utilizing different trading models and tariff structures.

vi. Provide RTPV owners more flexibility, further incentivizing uptake of DERs in UP.

vii. Reduce the financial impact of net metered solar systems on MVVNL.

viii. Testing of automated trading logics at customer end, which will help in removing manual intervention in the trading, especially with TOD/TOU tariff regime that was enabled by setting of minimum/maximum dynamic pricing.

ix. Support acceleration of deployment of RTPV through cutting-edge technology–supported market-based mechanisms.

x. Provide valuable opportunities to UPPCL to:

- Learn how best to implement a network tariff to support the wider rollout of P2P electricity trading

- Understand the impacts of P2P trading on the electricity distribution network

9.7 PROJECT IMPLEMENTATION DETAILS

In order to prevent service disruptions, ISGF installed smart meters in 12 buildings in series with the existing MVVNL's meters. A local firm, Technosun India Pvt Ltd in Lucknow, engaged in RTPV installations, assisted in seamless installation of the smart meters procured from Crystal Power. Vodafone's 4G SIM cards installed in the smart meters communicate the energy flow data through the head-end system (HES) of Crystal Power, which is integrated with the Blockchain platform of Power Ledger. The hardware connectivity diagram before and after installation of new smart meters is depicted in Figure 9.1.

The ISGF team collaborated with Abjayon Inc. to integrate the Power Ledger's Blockchain platform with UPPCL's customer care and billing (CC&B) system. To ensure reliability and quality supply and uninterrupted billing process during the project phase in UPPCL, only mock trading was conducted, on a separate instance of CC&B environment

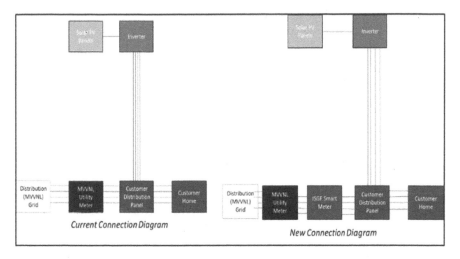

FIGURE 9.1   Hardware connectivity diagram.

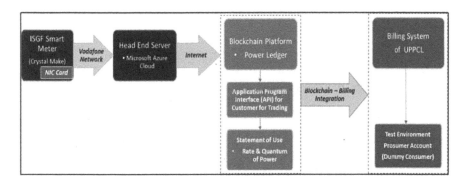

FIGURE 9.2   Information flow diagram from meters to billing system of UPPCL.

created on the cloud. Figure 9.2 depicts the information flow from the meter to the billing system of UPPCL.

9.8 P2P TRADING OPTIONS—EXPERIMENTAL

9.8.1 Fixed-Price Trading

This ensures P2P trading at a fixed price and guarantees each user certainty over the price they will receive for their energy traded through P2P. If no excess energy is available in a P2P network, then the consumer's energy needs will be met by DISCOM. Similarly, if no buyer is available for the P2P, then it will be sold back to DISCOM. A pictorial representation of the trade can be seen in Figure 9.3.

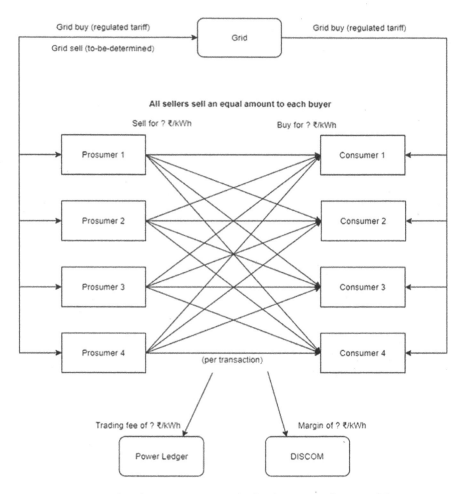

FIGURE 9.3 Graphical representation of a fixed-price trading model.

9.8.2 Dynamic Price Trading

The dynamic trading option involves prosumers and consumers trading with each other, setting their own prices. There can be numerous ways of finalizing the trade, which may be decided as per the regulatory environment/framework. The cleared price might be the highest price the buyer is willing to pay or the lowest amount the seller is willing to accept. A pictorial representation of the trade is shown in Figure 9.4.

9.8.3 Dynamic Price Trading with Preferential Trading

In this case, the prosumer can also be given a choice to identify its preferred consumer—the practice termed *preferential trading*. This rule

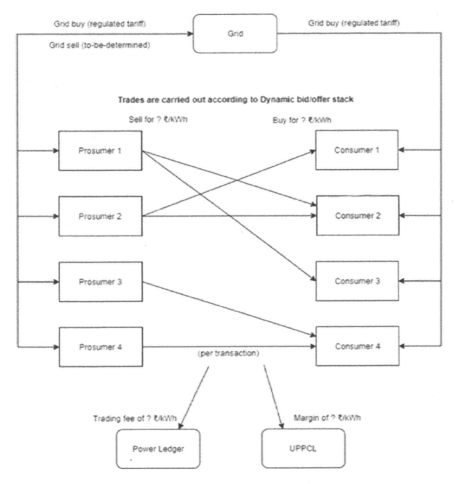

FIGURE 9.4   Graphical representation of dynamic price trading model.

allows prosumers to choose a consumer and offer them a percentage of their excess energy at a specific price, or any other mutually negotiated tariff that trade will be carried out before any other trading occurs, meaning that prosumers can choose their preferred off-taker. A pictorial representation of the trade is shown in Figure 9.5.

9.9 RESULTS

DISCOMs should enhance their metering infrastructure in order to allow energy exchange by installing advanced meters and developing systems for gathering and transmitting metering data to third parties. The availability of individual customer metering data is not only essential for allowing energy trading in energy markets, but it will also benefit all stakeholders

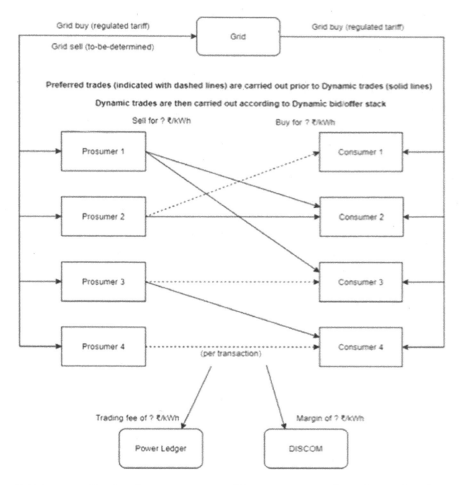

FIGURE 9.5 Graphical representation of dynamic price with preferential trading model.

in the system. Due to access to granular usage data, DISCOMs will have a much better understanding of the network and what improvements are taking place. Stakeholders may use this information to make better business decisions. This data may also be made accessible to third-party participants and service providers, enabling them to define areas for innovation and business opportunities in the residential sector.

The project went live on December 17, 2020, and the summary of three months of P2P trading executed on the platform is presented below:

a) Cumulative Community Sell (without P2P Trading): 7,694.42 units at a cost of Rs.14,186.72 at an average rate of Rs1.84/kWh. Even though

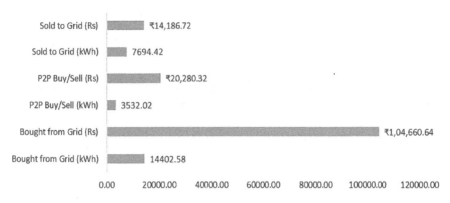

FIGURE 9.6   Community trading highlights—cumulative.

the feed-in-tariff (FIT) states that the surplus solar energy fed back into the grid is Rs.2/kWh, as per our observations, on an average, the real prices reflected is Rs1.84/kWh. This is the situation for transactions between prosumers and utilities. The prices are too low to be appealing for investments from the perspective of the prosumer/consumer, and thus serve as a barrier to rooftop solar adoption in the state.

b) Cumulative Community Buy (without P2P Trading): 14,402.58 units bought from grid at a cost of Rs.1,04,660.64 at an average of Rs.7.27/kWh. With an average connected load of 5 kW, the community's energy consumption has been rising with the arrival of the summer months, as shown by the change in the quantity of energy purchased from the grid during the winter months (December 2020 – January 2021) versus the arrival of the summer (January 2021 – February 2021).

c) Cumulative P2P Transactions: 3,532.02 units at a total cost of Rs.20,280.32 at an average cost of Rs.5.74/kWh. When transactions are P2P, the average price clearly represents the per-unit energy savings of Rs.1.53 (Rs.7.27 – Rs.5.74). With proper accounting and billing, these transactions are safe and secure.

Figure 9.6 depicts the cumulative community trading highlights. Based on the results from the dashboard, we have seen an average grid sell price of Rs.1.84/kWh, grid buy price of Rs.7.27/kWh, and P2P price of Rs.5.74/kWh. The graph in Figure 9.7 depicts the monthly average prices (Rs/kWh).

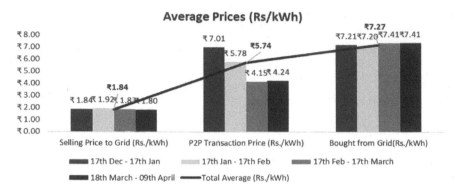

FIGURE 9.7   Average prices (Rs/kWh).

9.10 RECOMMENDATIONS

With the multitude of schemes and incentives offered by national and state governments to promote rooftop solar, the number of prosumers and solar generation capacity are going to increase in near future in Uttar Pradesh. Enabling P2P energy trading would provide a platform for both prosumers and consumers to buy and sell green energy under the rules and regulations that may be defined by the Honorable UPERC, enabling them to better recover their costs and increase green energy in the state. It would further help in achieving Sustainable Development Goals 7 ("Ensure access to affordable, reliable, sustainable and modern energy for all") that India is committed to [6].

In view of this, ISGF, in consultation with Power Ledger and the Project Advisory Committee, proposes the following path forward:

i. P2P Trading Price: During our interactions with the project participants, the majority of prosumers/consumers said they do not wish to bid/trade daily. Rather, they would opt for a fixed price for sale and purchase of RTPV energy, which would give them better clarity on how much higher price they can sell (for a prosumer) and save (for a consumer), if they engage in P2P trading. In view of the project experience, we recommend keeping P2P sale price of RTPV energy at mid-point between the highest commercial tariff and the net-metering feed-in-tariff. These prices may be kept the same for a financial year. When tariff regulations are issued, the P2P trading price can also be issued by UPERC.

ii. Fee for the DISCOM: In theory, the DISCOM revenue will be reduced if customers are buying electricity from each other. However, the total RTPV energy produced in the state being well below 0.1 percent of the total energy sold by UPPCL, the impact on their revenue will be negligible in the near term. Since DISCOM's network is used for the power flows and their IT systems are leveraged for billing and settlement, DISCOMs are eligible for a modest fee that they may term as network access charges or service charges. To begin with, we recommend 15% of the sale price of RTPV energy as DISCOM fee, which may be borne by the buyer. For example, in the present scenario, the price to be paid by UPPCL for net export of RTPV energy is Rs2/kWh to the prosumer, whereas a commercial consumer pays the tariff of Rs8.75/kWh (for consumption above 1,000 kWh/month). The midpoint suggested for P2P trade of RTPV is Rs5.38/kWh. The buyer will pay the 15% of 5.38, which is Rs.0.81/kWh, to the DISCOM as a network access charge. In this scenario, the prosumer will have a net gain of Rs3.30/kWh for the energy they sell and the consumer who is buying from a prosumer will save Rs2.56/kWh.

iii. Achievement of RPO Targets by DISCOMs: Under Section 86 subsection (1) clause (e) of the Electricity Act 2003, RPOs mandate DISCOMs, open-access consumers, and captive consumers purchase a portion of their electricity from renewable sources [7]. If all the RTPV energy traded is eligible for RPO of the DISCOM, that is an added benefit for promotion of P2P trading among their customers. This will further reduce the DISCOMs obligation to buy renewable energy certificates (REC) from the market to meet RPO.

iv. MVVNL suggested to extend the P2P trading platform to agricultural feeders that are being solarized under the KUSUM scheme [8]: Under the provisions of the PM-KUSUM program, with a 30% each subsidy from the central and state governments and a 40% farmer's contribution, grid-connected agriculture pumps can be solarized. The solar capacity allowed is up to two times the pump capacity in kW, and DISCOM would purchase the surplus power. Although agriculture feeders have already been separated, the feeders with mixed loads having irrigation pump (IP) sets for agriculture can also be considered for solarization initially. Feeders may be selected based on load, technical and commercial losses, number of consumers, etc. The solar PV systems installed for the IP sets can be fitted with smart meters and connected to the Blockchain platform. However, the IP set customers on agricultural feeders may not be covered under the existing billing system of UPPCL with which the Blockchain platform is integrated. Those accounts may be ported to the CC&B system by UPPCL.

v. Regulatory Interventions: UPERC may issue formal trading rules and regulations. Open-access regulations currently limit customer and prosumer involvement in trading to those with 1 MW or more demand. As per existing open-access regulations, producers or users must apply to respective state load dispatch centers (SLDC) for permission to wheel traded power on the electricity network. There is no special provision for such P2P sales and purchases. Consequently, there are various types of charges that can be placed on either the seller or the buyer. For P2P trading through Blockchain, new clauses must be added to the current regulations, or a separate regulation may be issued.

vi. Capacity building must be a central component of sector reform implementation. Training is required for all stakeholders, including SLDCs, SERCs, and DISCOMs, as well as for unique customer groups. Nongovernmental organizations and think-tanks can help by first explaining the benefits to stakeholders and then providing capacity-building tools to aid change preparation and implementation.

9.11 CONCLUSION AND WAY FORWARD

To take the technology beyond the pilot stage sandboxes in India, a set of regulations and technical specifications for service must be agreed on and followed by all stakeholders in the P2P network. To accomplish this, regulatory environments in the power sector in various states must take the lead in clearly identifying stable regulatory structures with frameworks that facilitate and promote the proliferation of decentralized transaction models based on regulations and beneficial practices observed across the world. As a result, when defining P2P energy trading rules to promote the proliferation of RTPV, the most important factor to consider is striking a good balance between the greater good for all by defining a collection of direct, consistent, and equitable regulations.

ACKNOWLEDGMENTS

I would like to record my appreciation and sincere thanks to all the people involved in this first-of-its-kind project in South Asia, and in India. First, my thanks go out to Shri RP Singh, Chairman, Uttar Pradesh Electricity Regulatory Commission (UPERC), for actively supporting such visionary regulations that enabled us to conduct the pilot in the first place. Second, I would like to thank the continuous and coordinated support received from the chairman, managing director, and other senior and junior members of Uttar Pradesh Power Corporation Ltd. (UPPCL) and the utility company, MVVNL. Abjayon Inc., the system integrator for UPPCL, deserves to be acknowledged as well for enabling a seamless integration of the Power Ledger's Blockchain platform with the existing UPPCL billing system. In addition, I would also like to thank the entire team at Crystal Power Energy Meter and Technosun India Pvt. Ltd., who despite numerous challenges encountered due to the COVID-19 pandemic ensured the timely supply, installation, and operation of the meters installed at the customer premises. I would also like to thank and acknowledge the entire team of Power Ledger, Australia, for partnering with India Smart Grid Forum (ISGF) and sharing their expertise and Blockchain-based platform that enabled the success of the pilot. It would certainly not have been possible without the guidance and constant mentoring of Shri Reji K Pillai, President, ISGF; Smt. Ms. Reena Suri, Executive Director, ISGF; and Shri Suddhasatta Kundu, Senior Manager – Technical Advisory, ISGF. Lastly, I thank and congratulate all the participants for the generous and helpful guidance they provided in ensuring and engaging in continuous participation throughout the project tenure.

REFERENCES

[1] "Blockchain Technologies for Smart Energy Systems: Fundamentals, Challenges and Solutions."
[2] "Grid Connected Solar Rooftop Programme, Ministry of New and Renewable Energy (MNRE)." https://bit.ly/3xDxEA1.
[3] "Programme/Scheme Wise Physical Progress in 2020-21 & Cumulative up to March 2021, Ministry of New and Renewable Energy (MNRE)." https://bit.ly/35PCGgZ.
[4] "Draft Electricity (Rights of Consumers) (Amendment) Rules, 2021; Sub-Rule (4) of Rule 11 of the Principal Rules." https://bit.ly/2TNMe9q.
[5] "UPERC (Rooftop Solar PV Grid Interactive Systems Gross/Net Metering) Regulations, 2019 (RSPV Regulations, 2019), Clause 10.4, Section (vii)." https://bit.ly/2SBNXOV.

[6] "United Nations SDG Goals 7: Ensure Access to Affordable, Reliable, Sustainable, and Modern Energy for All." https://bit.ly/3wF53du.

[7] "The Electricity Act, 2003." https://bit.ly/3xv2XwE.

[8] "Pradhan Mantri Kisan Urja Suraksha Evem Utthan Mahabhiyan (PM KUSUM) Scheme, Ministry of New and Renewable Energy (MNRE)." https://bit.ly/3cQY0X4.

CHAPTER 10

A Framework for Blockchain-, AI-, and IoT-Driven Smart and Secure New-Generation Agriculture

Rohit Kumar Kasera, Raktim Deb, and Tapodhir Acharjee

Triguna Sen School of Technology, Assam University, Silchar, Assam, India

CONTENTS

DOI: 10.1201/9781003188247-10

10.1 INTRODUCTION

In recent years the agricultural sector is increasingly utilizing the services of technologies such as Internet of Things (IoT), robotics, machine learning (ML), Big Data (BD), etc. to optimize the entire value chain starting from input suppliers and up to end consumers [1, 2]. All these technologies are making the agricultural sector smarter and more efficient. Smart agriculture refers to the use of technologies like the Internet of Things (IoT), artificial intelligence (AI), robotics, and many others to improve both pre- and post-harvesting experiences. The ultimate goal is to boost the quality and quantity of crops and deliver best-quality and safe foods to the end-consumers utilizing the optimal human labor in the process. It is expected that the global smart agriculture market will reach $15.3 billion by the end of 2025 [3]. Today, all the pre- and post-harvesting activities in agriculture can be brought under the umbrella of smart agriculture. Pre-harvesting activities include preparing the soil by land tilling, using fertilizers, watering etc. in preparation for planting, then keeping constant care of the plants, by watering, pruning, etc. The post-harvesting activities include mainly agricultural product business such as handling of crops, storage, processing, packaging, logistics, food safety, etc. Post-harvesting activities are performed by not only farmers but also middlemen, from small businesses to large corporate entities, transportation people, and different government agencies to purchase and sell crops, ensure food safety, etc. and ultimately retail sellers and consumers. Some of the activities of smart agriculture include, among others, precision farming, livestock monitoring, smart greenhouse, smart forestry and horticulture, smart water management, yield monitoring, field mapping, crop scouting, farm labor management, weather tracking and forecasting, water quality management, irrigation management, and agricultural asset management.

People around the world are becoming more and more aware of the specifics about their food products like the time of harvest, place or condition of production, storage, logistics, and packaging [4, 5]. Aside from that, Food and Agriculture Organization of the United Nations and other researchers have mentioned the global food wastage statistics due to lack of planned production and distribution of food [6–8]. Therefore, security and transparency are also being given utmost importance in different pre- and post-harvesting activities. Agricultural crops or are susceptible to different kinds of contamination, intentional and otherwise. There should be some transparency through which all the stakeholders like farmers, warehouse manager, logistics people, big business houses, retail sellers, and customers of the agricultural products can feel confident about the quality of the product. Blockchain is an emerging technology that is providing the next level of security and transparency for different kinds of modern applications, including smart agriculture [9]. A Blockchain is a distributed ledger that is structured into a linked list of blocks. In each there is an ordered set of transactions. A ledger is distributed across many machines and contains stores of transactions that can only be appended;the entries in this ledger database are permanent and cannot be changed or altered. Any new transaction will be reflected on the distributed databases hosted on different nodes. The Blockchain components can be divided into five layers, namely application, execution, semantic, propagation, and consensus layers [10]. Various kinds of transactions in the network can be secured using Blockchain: for example, transaction of payments, online voting, making reservations of hotel rooms or flights, signing contracts, etc. We can exchange and track the records on the Blockchain platform, and all the participating agents can check the shared duplicate transactions records in the network [11]. Some of the applications of Blockchain in agriculture are secured data storage, remote monitoring and automation, storage of agricultural and environmental monitoring data, transparency and monitoring in agricultural trade, etc. [12]. Blockchains can be used in agricultural supply chains; one specific example being the tracking of food certificates and real-time updating of the certificates [13].

As per [14], IoT can be defined as interconnecting physical devices that allow data collection and exchange of information collected by each. IoT is now combined with other technologies like wireless sensor networks (WSN) and Blockchain to enhance the capacity and efficiency of different application areas like smart agriculture [15], smart grids, smart cities, smart homes, etc. [16]. Machine learning (ML) is another heavily used technology that is enhancing the capacity and efficiency of modern agricultural activities [17]. ML usually refers to the changes in systems that perform tasks associated with artificial intelligence (AI). These tasks could be recognition, diagnosis, planning, robot control, prediction, etc. AI is used in almost all pre- and post-harvesting activities today, from preparing the soil for seed, then planting a seed into the soil, and ending with the picking up of the ripe harvests by robots with the help of computer vision [18]. Some of the specific tasks for which ML is used in agriculture are: creating better crops seeds to sell to customers; identification of bacteria and pests and information sharing with pest control companies; suggestions for farmers about matching the type of cropsto the type of soil to produce the best yield; and suggestions on which pesticides to use depending on the crops type, pest type, and agricultural season [19].

In this chapter we deal with utilization of Blockchain in pre- and post-harvesting activities along with AI and Internet of Things (IoT) together. There are various existing research works showcasing the use of Blockchain, IoT, or ML in agriculture. But very few of them are combining these three technologies together for smart agriculture. In our proposed methodology, we have tried to utilize these three technologies together for more efficient smart agriculture. We have used Blockchain technology for storing information related to food and crops production so that farmers can easily access all information with a single smartphone. The gateway of this IoT-based food and crops production is designed module-wise. IoT sensors sense various changes in the target environment and record the values; then collected sensor data are cleaned using a normalization method; next, data are organized and trained for gaining additional and smart knowledge using the ML algorithms; and last, all predicted information is stored in the Blockchain network for access by the relevant parties. Blockchain will ensure transparency in every step starting from production of crops with quality of seeds and proceeding to fertilizers and pesticides used, warehousing details, logistics details, etc.

In summary, the contributions of this chapter are as follows:

1. Investigation of existing methodologies related to the use of Blockchain in today's smart agriculture. Some of these methods are also using IoT and AI. We discuss the strengths, weaknesses, and contributions of these existing methods.

2. We propose an architecture for smart agriculture using the combination of Blockchain, IoT, and AI.

3. We compare our architecture with other existing methodologies.

Section 10.2 presents a review of research literature related to the use of Blockchain, IoT, and ML in agriculture. In Section 10.3 a methodology is proposed for food crops productions using these three technologies. Next, in Section 10.4 we compare our methodology with other methods utilized in this area. Section 10.5 offers some conclusions.

10.2 LITERATURE SURVEY

10.2.1 Blockchain-Inspired RFID-Based Information Architecture for Food Supply Chain

The authors in [20] proposed a secured and transparent food supply chain using IoT and Blockchain. This method uses radio-frequency identification (RFID)-based sensor and Blockchain at the physical and cyber layers, respectively. Here a management hub is used to cater to the need of data storage and computation tasks, and through this the IoT devices can interact with the Blockchain. With RFID and blockchain technology used in the supply chain, the food product is visible to different entities in the chain. Within the logistics chain from storage to customer, the RFID-tagged food products can be easily tracked, and for each transaction a data block is created that contains the information about a particular package. After verification of the transaction by the entities, a block is created and appended to the chain of blocks. Some of the steps of the entire methodology include: allocation of address to the SensorIDs, defining the transaction information structure, transaction verification process, consensus between the terminals, block formation, and data storage. At the time of scanning the food products, the real-time data captured are updated in a Blockchain, which builds an immutable digital history. To find the information about a food package, any consumer or retailer can check and verify the data from the public ledger.

Salient features of the proposed methodology:

1. Easy and trustworthy tracking and verification of food products throughout the food supply chain.

2. Identify if there are any hindrances in the whole supply chain.

3. Any possible prohibited adulteration of food products is identified.

4. Reduces the waste of food products by determining the shelf life.

5. Transparent information is provided to all the stakeholders.

6. Allows recalls to be specific.

Some weaknesses that we have observed in this method are:

1. The work only demonstrates single sensor integration, but more parameters like moisture, light, or specific volatiles could be also measured using other sensors, also here

2. The IoT devices do not interact with the blockchain network through management hub, so it is unclear if there is any chance of breach of security possible in the management hub.

10.2.2 Internet of Things for Smart Precision Agriculture and Farming in Rural Areas

In [21] the authors discuss the use of smart agriculture using IoT network solution for farms and agriculture in rural areas. The method usesWiFi-based long-distance(WiLD) network, Fog computing, and wireless sensor networks (WSN)-based systems for a range of agricultural activities like irrigation, farm monitoring, optimum use of fertilizer, soil health testing, intruder detection, water quality monitoring, etc. Cross-layer medium access control (MAC) protocol and routing protocol are proposed along with the architecture and showing excellent performances in terms of delay and throughput. These protocols can adjust the duty cycle of the sensors depending on the traffic nature in the multi-hop IoT proposed. They have also evaluated the testbed thoroughly.

The architecture combines multiple networks for scalability and larger range of coverage. The proposed architecture uses a WiLD network and a set of IPv6 over low-power wireless personal area networks (6LoWPAN)-enabled WSN networks. The goal is to make the network cover more area and be more scalable. A gateway is used for providing security, identification, and sending and receiving of data between two networks. The gateway is compatible with IEEE 802.15.4 network (6LoWPAN) and 802.11 networks (WiLD networks). For connecting the rural remote location to the internet, the WiLD network is used. Also the method uses Fog computing for real-time control, analyzes data in milliseconds, and sends a particular period's calculated aggregated data, thus reducing latency. Application layer protocol used in IoT-like message queuing telemetry transport (MQTT) and constrained application protocol (CoAP) are used in this method for receiving sensor data. Lastly, a cloud in IoT is used to provide the facility with computing resources from the storage server in on-demand fashion.

Salient features of the proposed methodology are:

1. The architecture and testbed are proposed and discussed in detail.

2. The testbed is also evaluated and produces the desired results.

3. The cross-layer MAC and routing protocols give excellent evaluation results.

4. The method is scalable and can connect remote nodes, as it uses a combination of technologies like WiLD, 6LoWPAN network, and Fog computing.

5. The Fog computing nodes save bandwidth and send the remote data with less delay.

6. Fog computing increases the agility of the system, as the developers can quickly develop cost-effective applications and deploy them whenever and wherever needed.

But there are some observed shortcomings of the method, as well:

1. Congestion in a larger network is not appropriately considered.

2. Bandwidth allocation in WiLd is static and thereby can lead to its inefficient use.

3. Security, data protection, and data contamination are of concern in this method.

10.2.3 Blockchain-Based Traceability in Agri-Food Supply Chain Management: A Practical Implementation

This study [22] proposes AgriBlock, a Blockchain-based decentralized system that uses Ethereum andHyper Ledger Sawtooth to trace the agri-food supply chain. Here the IoT devices sense the status of the food package and store the digital values in the Blockchain. If certain conditions are satisfied in the sensed data, a smart contract is triggered. In the study, both networks (Hyper Ledger Sawtooth and Ethereum) were configured with the default settings. Here it was observed that Hyper Ledger Sawtooth outperforms the Ethereum Blockchain: Ethereum latency is 16.55 seconds, network txis 528108 bytes and CPU load is 46.78%, whereas Sawtooth latency is 16.55 seconds, network tx is 19303, and its CPU load is 6.75%.

The process of supply chain management is the one in which the provider supplies the raw materials, producer is the farmer, processor is the one responsible for packing or processing of the produced items, distributors are responsible for distribution of the processed items to the retailers, retailers are responsible for selling the finished products, and consumer is usually the one who buys the items.

Representational state transfer (REST)-oriented architecture is used as an application programming interface, which strongly demonstrates the capabilities of the AgriBlockIoT to other applications and at the same time can allow smooth integration with other existing software. To convert the function calls of the Blockchain layer from high level to low level and vice versa, a controller component is proposed here. Smart contracts implement all the business logic of the Blockchain and are used as the gateway to the Blockchain itself.

Salient features of the proposed methodology are:

1. Due to the use of Blockchain and IoT, all the benefits of these technologies are nicely exploited by the system.

2. Use of Ethereum enhances the scalability and reliability of the system, as it can accommodate more entitiesin the system.

3. The Hyper Ledger Sawtooth provide the concept of consensus algorithm for this application and devices used.

Some observed weaknesses of this method are:

1. Ethereum is a costly network to maintain.

2. Even if we use private networks in Ethereum, including more sophisticated business logic will be troublesome due to the limitations of having only one language for programming smartcontracts and a fixed structure for the records.

3. The computation facility needed for calculating the consensus in Ethereum is huge, which may lead to some implementation challenges in the edge gateways and IoT devices.

10.2.4 Blockchain- and IoT-Based Food Traceability for Smart Agriculture

This study [23] mainly reflects on a food traceability system using recent technology like Blockchain and IoT. It is mentioned that some individuals were even concerned that all animal food would include hormones, while all plant food would contain poisons and colorants. It also highlighted several issues with today's food production, supply chain, and processing, and causes for all these issues are well defined in the Introduction part of the paper. The reasons for these issues are classified into four categories and explained in detail. While the Introduction does present an explanation of benefits, IoT framework and Blockchain for smart agriculture are well defined in the literature of the paper; however, no citations are mentioned in the first paragraph of "Related work" section. Also, a smaller number of paper citations are observed in literature review. The method this study proposes offers a trustworthy, self-organized, open, and ecological food traceability system that engages all partners in a smart agriculture ecosystem using Blockchain and IoT technology. The study does not, however, offer any implementation and experimental results—only a theoretical overview with a block diagram as the solution to the problem.

10.2.5 Blockchain and Smart Contract for IoT-Enabled Smart Agriculture

This study [24] proposes a comprehensive system using IoT and Blockchain for different pre- and post-harvesting activities. The IoT sensors are used to sense the quality and condition of the stored agricultural products. The data about pricing of the agricultural products and services are also monitored and shared by the IoT devices. Blockchain with smart contract makes sure that the data are secure and transparent to the authorized entities of the system. MQTT server is used to store the sensed data while Blockchain stores the more sophisticated data. Several actors are part of the system, namely contract owners, seed storage sites, supply shops, producers, distributors, wholesalers, retailers, and consumers. The implementation of the system is done through Remix web-based IDE, which provides an Ethereum wallet with dummy ether cryptocurrency. Blockchain continuously keeps track of the important data in pre-harvesting activities, such as quality of seed, and executes self-checks if any parameter such as temperature, humidity, light, etc. is violated. The same kind of transparent system is available to the authorized entities for keeping track of the seeds in the post-harvesting phase. Also the Blockchain keeps track of the data in each step of distribution right from the agricultural field to the customers. The authorized customers can also verify the data from the Blockchain.

There are many observed strengths of the method, such as:

1. It is a complete system where pre-harvesting and post-harvesting activities are brought under the umbrella of Blockchain.

2. IoT devices are used to sense real-time data and trigger the necessary function if any violation in the sensed data is found.

3. All the benefits of the Blockchain are nicely used in the system.

4. The system is tested with remix.

Some of the system's observed shortcomings are:

1. It does not use ML algorithms.

2. The system is yet to be tested in real Blockchain.

3. The IoT devices also need to be made secure.

10.2.6 Blockchain Technology for Agriculture: Applications and Rationale

This study [25] mainly focuses on the applications of Blockchain technology for agriculture. The paper provides experimental results, particularly in food supply chains, agricultural insurance, smart farming, and transactions involving agricultural products. As the abstract indicates, the study is to cover the practical perspective on the applications of Blockchain technology, but there is no result analysis or discussion section present in the paper. There is a mention of the experimental analysis on food supply chains, agricultural insurance, smart farming, transactions involving agricultural products, etc., and the challenges of recording transactions made by smallholder farmers and creating an ecosystem for utilizing the Blockchain technology in the food and agricultural sector are well defined in the paper.

10.2.7 BlockChain with IoT, an Emergent Routing Scheme for Smart Agriculture

The authors of [26] offer an effective routing strategy for distributed nodes that function in a distributed way to make efficient use of communication lines by combining IoT and Blockchain. The study also investigated the potential of a Blockchain to manage IoT devices. It shows how the dispersed peer-to-peer network among controllers leads to a stable and similar design of the suggested architecture. To identify a path to the base station, the proposed protocol employs smart contracts in heterogeneous IoT networks. Each node may guarantee a path from an IoT node to a sink and subsequently to the base station, allowing IoT devices to work together during transmission. The proposed routing protocol eliminates duplicate data and prevents IoT architectural attacks, resulting in lower energy consumption and longer network life. The result sare compared between low-energy adaptive clustering hierarchy (LEACH) in agriculture and the proposed IoT-based agriculture approach. In future research, a smart model based on IoT and Blockchain is likely to be created for clustered farm environment monitoring and information sharing with farmers and other stakeholders in order to make timely decisions to increase agricultural productivity. Overall, an energy-efficient IoT with Blockchain scheme is presented in this paper, and the simulation results demonstrate that this method has a longer network life, uses less energy, and has better throughput than LEACH in agriculture.

10.2.8 Study of IoT Blockchain Used in Smart Agriculture for Notification Well-Being and Preservation

This study [27] shows the implementation of smart agriculture using IoT Blockchain. In a Blockchain, nodes obtain the data from sensors. The sensors are connected to different things that are related to agriculture for monitoring. The node elements of IoT Blockchain used in smart agriculture include temperature sensor node, pressure sensor node, pressure control node, pollution control node, moisture water control node, smoke, fire control node, PH control node, etc. Every node in the Blockchain network works as a miner and has a local duplicate of the Blockchain including all confirmed activities. The activities that are performed in every node are retrieving, storing, and observing sensor information. In this paper, a Smoke Fire Control Node has been discussed using a single node. Operations performed include smoke fire control activities, reserve smoke fire details activities, acquire the smoke fire details activities, and observe the smoke fire situation activities. The system has been developed using an Ethereum private Blockchain network that supports the genesis block. An Ethereum account is externally controlled, and a liability account is regulated by a record code. In this fire smoke system, all nods are in an externally recognized account and its whole smart agreement is placed in the agreement account of observing agriculture. Whenever the block code is accomplished, the agreement account gets a message. IoT devices are programmed by a LibCoAP library of COAP in C. This creates a private/public key for each device from which the IoT devices are identified. To examine the system achievement a benchmark tool is applied, which is used to transfer the validated applications and wait for the acknowledgment. The handling of the user interface is done in java script from which IoT device programs are created to operate with the Blockchain network. It accepts the application and retrieves the data from the Blockchain and acknowledges the IoT device.

10.2.9 Supply Chain Management in Agriculture Using Blockchain and IoT

In this paper [28] the authors have proposed a Blockchain-based web-based platform for agriculture supply chain management (SCM) called FARMAR, which keeps track of all the points in the supply chain with IoT devices and uploads all the data in the Blockchain. In this method they have used a BigchainDB, which is a database with properties of Blockchain, such as decentralized control, immutable data storage, high throughput, powerful query operations, etc. In BigchainDB data is structured using various databases like SQL, NoSQL, JSON, key values, and tables. They have also used other technologies such as Monit and MongoDB for implementation. Transactions, assets, metadata, blocks, etc. are all stored in the BigchainDB. Three steps are there for submitting a transaction to BigchainDB, which are transaction payload preparation, its fulfillment, and uploading transactions using HTTPS.

The proposed method is providing the farmers a tool for bypassing the middleman in the system; ultimately the agricultural products can reach the customer in a more cost-efficient way. But there are some issues that are not taken into consideration or are not discussed, like the place off installation of the IoT devices, communication protocol used, security of the devices, and the efficiency of the method.

10.2.10 An Agri-food Supply Chain Traceability System for China Based on RFID and Blockchain Technology

In this study [29] the author has discussed the use of RFID and Blockchain in agri-food supply chain in China. The RFID is used for collecting and sharing data for production, warehousing, sales and supply, etc. The method uses Blockchain technology to safeguard the information shared and published in the traceability system. Here the traceability system covers authorized private parties like retailers, government organizations who are entitled for supervising the food safety and quality, and third-party regulators. The system would be very helpful for the government regulators, as they can check at any time the food packages, and if any hazardous issue is found, they can take appropriate measures to stop the distribution. Geographic information system (GIS) is also used here to track the crops. Global positioning system (GPS) is also used in this method for real-time tracking of distribution vehicles and to find the optimum route for the vehicles to reach the destination.

The RFID tags in the agricultural products carry useful information such as name of the crop/plant product, planting time, production area, use of fertilizer and pesticides, harvesting time, etc. This method can also be used for RFID tagging of meat product and carry information such as parents of the animal, fodder used for feeding, checking schedule for epidemic prevention, use of medications, etc. The data about production managers and operational staff are also uploaded in the system so that they can be easily tracked and contacted whenever needed. Through wireless network all the information of RFID tags is uploaded in Blockchain.

Then a processing enterprise scans the food products received from the production enterprise, and the RFID tags are updated with current status. Accordingly, the Blockchain is also updated. Next, the warehouse management link is used to track information about the condition of the food product and climatic conditions inside the warehouse. Again, the information is uploaded in the Blockchain. To ensure the 3T principle (time, temperature, and tolerance) for the food products, temperature and humidity sensors are used in the vehicle, the data is uploaded in real time for everyone's verification, and, if required, the necessary steps are taken to safeguard the products. Ultimately when the authorized retail seller or customer want to verify the freshness, quality, and other parameters of the product, they can easily do so through sales link information tracing.

Salient features of the proposed methodology are:

1. With this method all the authorized agencies, both private and government, can check and verify the quality and safety of the food product

2. GIS is used to track the condition of crops.

3. RFIDs in food product ensure the originality of the product.

Some observed shortcomings are:

1. The cost to be incurred will be high, which will translate into the higher retail price.

2. Storing of blocks of the Blockchain will require a lot of storage.

3. Device security is not taken into consideration.

10.2.11 A Framework for Blockchain-Based Secure Smart Greenhouse Farming

In this study [30] a lightweight Blockchain-based architecture is proposed for security and privacy in smart greenhouse farms. They have considered four groups, namely smart greenhouse, overlay network, cloud storage, and end user, for their system model. The smart greenhouse has various IoT devices such as different sensors for light, humidity, CO_2, water level, etc. and different actuators such as LED light, fan, heater, sprinkler, etc. A centrally managed Blockchain is proposed through which the devices can share the data securely and transparently. A policy header is added in every block of the Blockchain that contains the updated policy. The owner can manage the transaction through the access control list present in the policy header. The overlay network here is comprised of clusters, and every cluster has a cluster head (CH). The CHs can be changed by the members depending on the delay in the network. CH must make the decision of keeping a new block or discarding it depending on the receiving transaction. The greenhouse data are also stored in the cloud so that experts can view and analyze the data and provide useful suggestions and services accordingly. A unique block number is assigned to each data block, and this unique block number and hash values are also used for authentication. The owner is the end-user here who can centrally manage the system with their own device. The authors have defined four layers here for security framework, namely physical layer, communication layer, database layer, and interface layer.

The strengths of the proposed method are that it is simple and light-weight architecture and may be useful for a small farm or greenhouse. But the authors have not cited any test results. Also, the architecture proposes different groups, and the tasks of different groups are mentioned. The finer issues like data structure, communication protocols used, application protocol used, etc. are not discussed in detail.

10.2.12 AgriTalk: IoT for Precision Soil Farming of Turmeric Cultivation

This study [31] proposes AgriTalk, an inexpensive IoT platform for precision farming of soil cultivation. This paper uses turmeric cultivation to demonstrate precision farming. The authors showcase the importance of using IoT for cultivation. In this proposed method an Arduino microcontroller board called as AgriCtls connects the IoT devices such as sensors and relays. All the operations of IoT devices such as configuration, accounting and reporting, analysis, dashboard controls, etc. are performed through smartphones by AgriGUI, which accesses the AgriTalk cloud-based server.

Various IoT devices such as soil and insect sensors, 3-in-1 moisture sensors, etc. are used in the field to gather data about climate conditions, growth of turmeric, detect insects, etc. and send them to the IoTtalk engine in the cloud. The collected data are used to predict the climatic condition suitable for precision farming. Actuators are used in AgriTalk for turmeric cultivation. Automatic pesticide sprayers are used for pest control. The flying insects are kept away using repellent bulbs. Drip irrigation is also used depending on climate underneath the soil.

Salient features of the proposed methodology are:

1. It is a cost-efficient method, as the devices used are not particularly costly.

2. It is environmentally friendly, as there is no emission of greenhouse gases.

3. Failure detection and calibration of sensors are easy.

4. Production quality can be significantly enhanced.

5. The users can configure the devices easily with required operation and intelligence.

6. The farmers can use smartphones for easy access through AgriTalk.

But we have also observed some shortcomings of the proposed method, which are:

1. Constant power source is required for all the sensors and IoT hardware to run seamlessly.

2. Wind speed and wind direction in field were not collected, which might be a cause of slow growth of the turmeric plants.

3. No ML algorithm was used to make the intelligent prediction.

4. The method was tried on just one type of plant, so there is a need to test it on other crops before wider implementation can be considered.

10.2.13 Blockchain-Based Commercial Applications for Pre-and Post-Harvesting Activities

There are many available Blockchain-based commercial solutions launched by many start-up companies and reputed organizations that are really transforming the agricultural industry. Some of these are:

1. Agrichain [32]: It provides a platform to all the stakeholders in supply chain management and allows them to make wise decisions, removes paperwork, increases supply chain efficiency, reduces risk, etc. Some of the services provided by them are stock management, contracting, tracking supply chain, automation of logistics, traceability, broker integration, etc.

2. Agridigital [33]: It is another platform that can be used by farmers, traders, and others to make agricultural transactions using Blockchain and smart contracts. They are providing simple, easy, and secure ways for buying, selling, moving, storing, and reporting of the grains.

3. Agriledger [34]: This platform verifies digital identity for all stakeholders, providing market information, traceability of food origins, financial services, record keeping, etc. in agricultural supply chain.

4. IBM Food Trust [35]: It provides authorized, permanent, and shared information to the entities in the food supply chain. With sophisticated Blockchain-based solutions they are definitely enhancing the experience of the participants in terms of food safety, freshness of food, reducing food fraud, minimizing wastage of food, and ultimately reducing supply chain inefficiencies.

5. Agri10x [36]: It is a start-up organization that helps remote farmers sell their crops directly in the market using their Blockchain and AI platform. They are providing efficient solutions to the farmers in pre-harvesting activities. In the post-harvesting process, they are assisting the farmers with selling their products by providing facilities like traceability and security, faster and more secure payments, etc. Buyers can also use the system to trace and check the quality of products. Using AI they are analyzing the current market data for providing farmers and buyers fair pricing on the agricultural products.

Blockchain-based solution by Agri10x has simplified the post-harvesting process. Offering features like traceability and data security, farmers are selling on Agri10x platform and receiving benefits like fast payment settlement. The buyers, on their end, are assured of receiving quality produce.

10.3 METHODOLOGY

An increasing global population in turn increases challenges related to food production including crops safety and supply process with limited resources. There should be some smart transparent decision system with the help of which this food production can be monitored in real time with a limited number of support mechanisms. The current challenges include producing more food with fewer support mechanisms, improving customer satisfaction, allowing transparency over the stock connection, improving the safety of farmers' income, examining weather changes, etc. Using the IoT withB lockchain, this problem can be resolved by implementing a set of nodes and creating a mesh topology environment to distribute publicly the status of food production. For making farming supportable and transparent, there should be an uncomplicated methodology from which farming supplies (water, energy, manure) can be optimized in an easy way. Using Blockchain in agriculture improves and increases the production of food and crops and to output successful results. By using Blockchain there will be the reduction of wastage, supportable business, smart decision-making, and secure transactions. One of the biggest challenges in modern agriculture is the lack of secure, effective connection between various stakeholder, from chemical companies producing fertilizer, to seed banks, to farmers, to manufacturers of agricultural equipment, to shippers, storage specialists, wholesalers, retailers, etc. Each industry group has its own rules within the chain operation as well as their information reporting system [37]. This understandably makes it very challenging to track the related information and to share it among stakeholders. Using IoT with Blockchain makes this process comfortable and secure. In the following, we present the design of a comprehensive architecture (Figure 10.1) as a smart agriculture system for food and crops production using IoT and Blockchain.

The architecture diagram in Figure 10.1discusses step by step the process for food and crops production.

10.3.1 Information Production through IoT Sensors

For collecting sensor data, IoT devices and sensors are used. A smart system is developed in this way so that monitoring can be in real time on the farming field, at the warehouse, and at the retail store. Sensors used include temperature (infrared thermometer's sensors), soil (volumetric sensors, tensiometers), humidity (capacitive, resistive), light (phototransistor), fire alarm, etc. The IoT network for wireless information transmission is IEEE 802.15.4 (Zigbee) long-range wireless area network (LoRaWAN) [38]. By using these IoT devices all data are received and organized before it is saved and stored.

- Temperature Sensors: Temperature sensors are used to estimate the volume of heat pressure and coldness produced by any object or operation. It can sense and identify any physical modification in

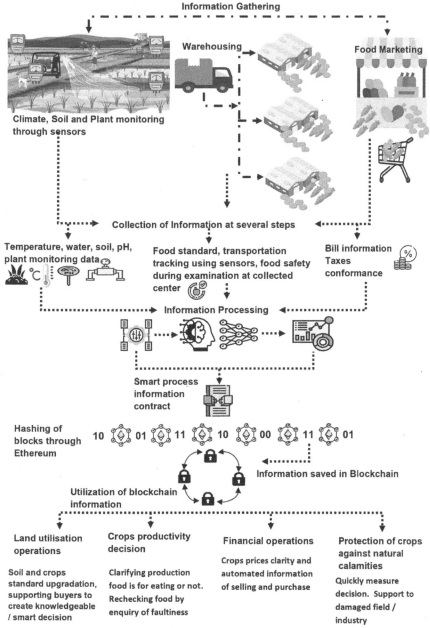

FIGURE 10.1 Proposed smart agriculture architecture using Blockchain, IoT, and machine learning.

the temperature and display it in an analog or digital signal. There are different types of sensors such as the ones for humidity, pressure, proximity, level, accelerometers, gyroscopes, and gas sensors. In agriculture, temperature sensors can monitor soil temperature, which is essential for effective crop growing.

- Water Quality Sensors: Water sensors are used to recognize water quality and monitor ion content in water circulation. Water sensors include chlorine residual sensor, total organic carbon sensor, turbidity sensor, conductivity sensor, pH sensor, and oxygen-reduction potential sensor.

- Other sensors used crops production include humidity sensors, motion detection sensors (to prevent theft), image sensors to compare the same crops at different time periods to detect disease, and smoke detectors to prevent fires.

- ZigBee: Zigbee is a wireless sensor network that has low energy, low data rate, low expense, and short-term suspension features. It is easy to deploy and extend. It provides strong protection and long-term data authenticity. Zigbee has mesh network nodes that use the transportation assistance of IEEE 802.15.4. Zigbee has logical device types such as coordinators, routers, end devices, and access nodes.

- LoRaWAN: This is a low-power wide area network that works on a low data rate. Variations of this technology include several standard networks such as LoRaWAN, SigFox, NB-IoT, and Weightless. It allows monitoring of facilities such as a warehouse in real time. It has a strong security mechanism from the device to the applications. LoRaWAN uses a star topology network standard, which as open-access protocol that works on top of a LoRa physical layer. Figure 10.2 shows the mesh topology architecture of LoRaWAN used to broadcast data.

Figure 10.2 shows that LoRaWAN can be used on multiple platforms such as indoors for warehousing and retail monitoring and outdoors in the farming fields for crops production monitoring. These platforms are interconnected with each other via the LoRaWAN gateway and transmit all data to a central gateway, which is an IP-based network. From this gateway, users access all event information through a LoRaWAN application.

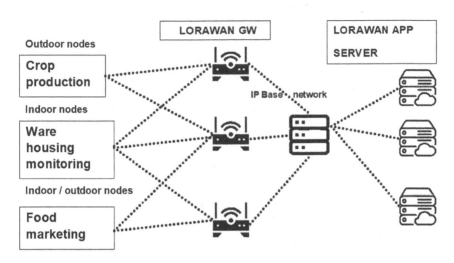

FIGURE 10.2   LoRaWAN mesh topology architecture for crops production.

10.3.2 Purification and Normalization of the Acquired Information

The sensors generate the information continuously, and as a result, some of the data may be repeated or irrelevant. To structure the data in a way that will be useful, it needs to be clean and in a correct format so that it easy to understand.

Once the cleaning operation is completed, all the relevant information is distributed at the information warehouse stage. Data should be constructed and well organized, and for this, some information should be included, namely a time stamp, demography, and which kind of data it is (string, number, float, etc.) while performing this operation. Storing data on the Blockchain network makes them more incorruptible and ensures that individually visible data connected with the data obtained from IoT tools are fully preserved and support the protection rules. Figure 10.3 presents the process of cleaning the data. First all sensor data from different stages are imported, and anything duplicated or irrelevant removed. Data are ready and reconditioned to get normalized. After the data are normalized, the testing is started to export the data for use. When the collected information is improved, it is put into ML models for training and testing.

10.3.3 Gathering More Specific Knowledge through Machine Learning

The cases that will be used for building the model are crop classification support, crop recognition, crop submit forecasting, crop spreading measurement score, and crop market request forecasting. Using the

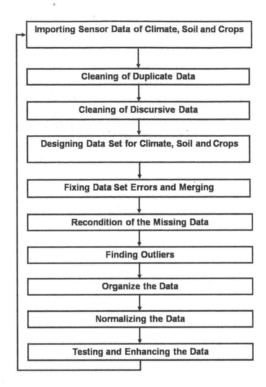

FIGURE 10.3   Data cleaning process.

ML algorithm of the captured data the farmers and stakeholders will be able to increase productivity and enhance the watering operations [39]. Storing data in the Blockchain network will allow agriculture contributors (farmers, suppliers, packagers, and resellers) to acquire such data explicitly. The flow diagram in Figure 10.4 shows how a ML algorithm will work for the crop production–related forecasting. For classifying several crops' temperature, soil-related data for predicting a data mining procedure are important. The analysis of this agriculture data using different data mining techniques is easily achieved. The algorithm includes K for clustering and fuzzy c for forecasting [40]. Using multilayer perceptron as deep learning neural network, a time series prediction is created for weather forecasting, selling of crops, and warehousing distribution. For performing overall production of food in industries it uses modern processing method based on genetic algorithm and statistical methodology as Bayesian network [41].

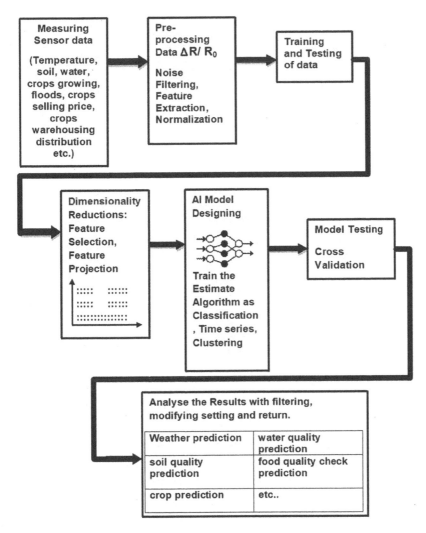

FIGURE 10.4   Process to train and test the data of food production using machine learning.

10.3.4 Storing Methodology

After applying an AI algorithm for forecasting food- and crops-related data, the information is saved in the Ethereum file system as a shared storage policy with hashing address on the blockchain network storage. To make the system more cost efficient, other Blockchain file system also can be used if and as required.

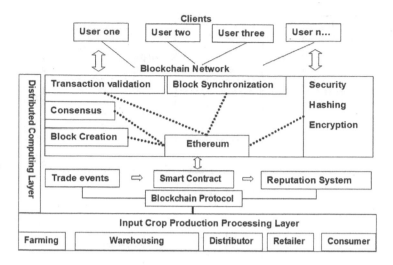

FIGURE 10.5   Storing and accessing of data in Blockchain.

With the use of a Blockchain network, the information is shared across each node in the network block as a peer-to-peer (P2P) connection, which means it does not need any central administration to manage the system. Figure 10.5 shows how the information is stored in the Blockchain network, and how the users participate in such a network. The data captured at different stages for food and crops production in the Blockchain can be accessed only after execution of a smart contract or managing controls established between them. Smart contracts support the interchange of information saved on each block among the network participants. This means data will be accessible to each agriculture store candidate associated with this network. In the distributed computing layer, all Blockchain-related operations will be performed, including transactions, consensus, security, and more. In the security section every transaction will have its hashing key, which is very difficult to decrypt. All transaction will be displayed as ledger in the Blockchain network, which will be available to all participants who are involved in this network.

10.4 COMPARATIVE ANALYSIS BETWEEN PROPOSED METHOD AND EXISTING METHODS

Table 10.1 presents a comparative analysis between the proposed method and the previously extant methods. The parameters compared include: the design system device; data protection, cost efficiency; intelligent system design; and the use of AI, IoT, and Blockchain. The analysis has determined that the proposed methodology supports ML, IoT, and Blockchain.

TABLE 10.1 Comparative Analysis between Proposed Method and Existing Methods

Reference	Methodology	Protected	LowCost	Intelligent System	Suuport AI/ML/IoT/Blockchain
Mondal, S., Wijewardena,K. P., Karuppuswami (2019) [20]	RFID-based food supply chain traceability using blockchain	Yes	Yes	No	AI/ML: No, IoT: No, Blockchain: Yes
Nurzaman Ahmed, Debashis De (2018) [21]	Smart wireless sensor network architecture designed for farming and agriculture using IoT Fog computing	Yes	Yes	No	AI/ML: No, IoT: Yes, Blockchain: No
M. P. Caro, M.S. Ali (2018) [22]	Traceability of food supply chain using Blockchain IoT	Yes	No	No	AI/ML: No, IoT: Yes, Blockchain: Yes
Pranto T.H., Noman A., Mahmud A (2021) [24]	Smart agriculture system using IoT and Blockchain	No	No	No	AI/ML: No, IoT: Yes, Block Chain: Yes
M.S hyamala Devi, R. Suguna, Aparna Shashikant Joshi (2019)[27]	Smart agriculture system design for safety and security using IoTBlockchain	Yes	No	No	AI/ML: No, IoT: Yes, Blockchain: No
Tian, F (2016) [29]	A RFID- and GIS-based system design for food supply chain traceability using Blockchain	No	No	No	AI/ML: No, IoT: No, Blockchain: Yes
Chen, W.L.; Lin, Y.B. ;Lin(2019)[31]	Soil cultivation system design using IoT	No	Yes	No	AI/ ML: No, IoT: Yes, Blockchain: No
Proposed Method	Crops production and food traceability usingIoT Blockchain and machine learning	Yes	Yes	Yes	AI/ ML: Yes, IoT: Yes, Blockchain: Yes

Comparing the proposed method with the extant ones shows that the proposed methodology is using IoT with Blockchain in which ML is being used in middleware to preprocess the sensor data so that imported data can be more knowledgeable and accurate. Table 10.2 shows that no other existing method can currently achieve the same.

TABLE 10.2 Comparative Analysis between Proposed Method and Existing Methods in the State of Technologies

Technologies/Methodology	Proposed Methodology	Previous Related Work by Mondal, S., Wijewardena, K. P., Karuppuswami in (2019) [20]
IoT sensors used for crops production: temperature, water, soil, fire alert	Yes	No
IoT data communication devices at the network layer as ZigBee, LoRaWAN for crops production, warehouse distribution, and food market	Yes	No
IoT devices, such as Bluetooth Beacon, used for food traceability and supply chain	Yes, Beacon Bluetooth is used with each food product to monitoring in real time food distribution from warehouse to consumer	No, RFID is used for supply of food
Network applicationused (LoRaWAN app server)	Yes	No
Artificial Intelligence used for making data knowledgeable and accurate of crops production, warehouse distribution, and food market	Yes, machine learning–based algorithms are used, such as clustering, Bayes' theorem, and time series prediction for making data filterable and knowledgeable	No such intelligence system has been used
Blockchain network for storing all data, distributed ledger of crops production, warehouse distribution, and food market.	Yes, Ethereum is used to create a smart contract with all farmers and consumers who are connected to this network. It is easy-to-use, secure technology. IoT devices data are linked with Blockchain enabling real-time monitoring. Machine learning is used as middleware for data filtration.	Blockchain network is used only for food supply chain but not linked with theIoT device. There is no clarity of which Blockchain network is used to transmit the RFID data.

Advantages of using proposed method:

1. **Building transparent administration and crop care**: with the use of Blockchain networks, the crops are better attended to. The detrimental effect of varying weather conditions can be reduced by deploying some IoT devices. Soil quality and fertilizer usage, among other factors, can also be better monitored, and combining IoT with Blockchain allows to do so with nothing more than a smartphone.

2. **Enhancing the productivity of the farmers**: Using Blockchain allows farmers to store all the information in one place, where it will be secure, well organized, and accessible at any time.

3. **Transparent and fast payments for farmers**: Currently, under the traditional systems, it may take farmers more than a week to receive payment for their agricultural products. Using Blockchain allows to speed up this process considerably through the use of smart contracts. If all the conditions of the transaction are satisfied, it is also completed without charging any special fees. Thus, farmers receive their payments quickly and their bottom line is less negatively affected.

4. **Enhanced food traceability**: Using IoT with Blockchain allows the products to be tracked easily and reducing the risk of fraud along the logistical chain. Tracking can be done along various steps of farming, collecting, warehousing, and distribution. With the help of Blockchain the information can be verified about source of food production, food transporters, food factories, retailers, and consumers. Trackable information includes the current status of food, current location of food, sender and receiver of food, the environment and climatic condition in the warehouse, preservatives used, etc.

10.5 CONCLUSION

This chapter began with a detailed discussion of how smart agriculture is being developed using IoT-based Blockchain. A literature review has been done of various proposed methods, with their strengths and shortcomings, such as the farming of turmeric, RFID-based architecture for food supply chain, smart precision agriculture and farming in rural areas, food traceability for smart agriculture, convenience analysis of sustainable agriculture based on Blockchain, and emergent routing scheme for smart agriculture using IoT. We also discussed some of the Blockchain-based commercial applications used in pre- and post-harvesting. Then, we proposed a smart agriculture system that works by information gathering regarding soil, plant, water, food warehousing, and food marketing through IoT sensors for monitoring. Using a machine learning algorithm and by training and testing the data using a neural network, this information becomes more relevant. After these operations, all the filtered information is stored in a Blockchain network for shared access by all authorized stakeholders connected through this network. The proposed methodology is compared with existing methods and found that our method is providing a comprehensive solution for different aspects of smart agriculture. The architecture remains to be implemented and tested in real Blockchain and analysis of results needs to be done in future.

REFERENCES

[1] Giorgia Bucci, Deborah Bentivoglio, and Adele Finco. Precision agriculture as a driver for sustainable farming systems: State of art in literature and research. *Calitatea*, 19(S1):114–121, 2018.

[2] Xuan Pham and Martin Stack. How data analytics is transforming agriculture. *Business Horizons*, 61(1):125–133, 2018.

[3] Global Smart Agriculture Market. https://www.zionmarketresearch.com/report/smart-agriculture-market, last accessed on 14.08.2021, online.

[4] 8 in 10 consumers check the origin of their food when purchasing products. https://www.newfoodmagazine.com/news/42541/8-10-consumers-labelling/, last accessed on 21.10.2021.

[5] More consumers focused on origin & quality of food. https://www.nutraceuticalsworld.com/issues/2019-03/view_breaking-news/more-consumers-focused-on-origin-quality-of-food/, last accessed on 21.10.2021.

[6] The State of Food and Agriculture. Food and agriculture organization of the United Nations, 2019, [online]. Available:ca6030en/ca6030en.pdf.

[7] Mónica Duque-Acevedo, Luis J. Belmonte-Urena, Francisco Joaquín Cortés-García, and Francisco Camacho-Ferre. Agricultural waste: Review of the evolution, approaches and perspectives on alternative uses. *Global Ecology and Conservation*, 22:e00902, 2020.

[8] Food and Agriculture Organization of the United Nations. Food wastage footprint full cost accounting,2014. [online]. Available: http://www.fao.org/3/a-i3991e.pdf.

[9] Xiwei Xu, Ingo Weber, and Mark Staples. *Architecture for blockchain applications*. Springer, 2019.

[10] Bikramaditya Singhal, Gautam Dhameja, and Priyansu Sekhar Panda. *Beginning blockchain: A beginner's guide to building blockchain solutions.* Springer, 2018.

[11] Olivier Alphand, Michele Amoretti, Timothy Claeys, Simone Dall'Asta, Andrzej Duda, Gianluigi Ferrari, Franck Rousseau, Bernard Tourancheau, Luca Veltri, and Francesco Zanichelli. Iotchain: A blockchain security architecture for the internet of things. In *2018 IEEE Wireless Communications and Networking Conference (WCNC)*, pages 1–6. IEEE, 2018.

[12] Oscar Bermeo-Almeida, Mario Cardenas-Rodriguez, Teresa Samaniego-Cobo, Enrique Ferruzola-Gómez, Roberto Cabezas-Cabezas, and William Bazán-Vera. Blockchain in agriculture: A systematic literature review. In *International Conference on Technologies and Innovation*, pages 44–56. Springer, 2018.

[13] Lan Ge, Christopher Brewster, Jacco Spek, Anton Smeenk, Jan Top, Frans van Diepen, Bob Klaase, Conny Graumans, and Marieke de Ruyter de Wildt. *Blockchain for agriculture and food: Findings from the pilot study.* Number 2017-112. Wageningen EconomicResearch, 2017.

[14] Vangelis Gazis. A survey of standards for machine-to-machine and the internet of things. *IEEE Communications Surveys & Tutorials*, 19(1):482–511, 2016.

[15] B.V.V.S. Narayana, K.S. Ravi, and N.V.K. Ramesh. A review on advanced crop field monitoring system in agriculture field through top notch sensors. *Journal of Advanced Research in Dynamical and Control Systems*, 10(6):1572–1578, 2018.

[16] Ana Reyna, Cristian Martín, Jaime Chen, Enrique Soler, and Manuel Díaz. On blockchain and its integration with IoT. Challenges and opportunities. *Future Generation Computer Systems*, 88:173–190, 2018.

[17] Nils J. Nilsson. *Introduction to machine learning. An early draft of a proposed textbook.* 1996.

[18] Machine learning in agriculture: Applications and techniques. https://www.kdnuggets.com/2019/05/machine-learning-agriculture-applications-techniques.html, last accessed on 28.08.2021.

[19] Role of machine learning in modern age agriculture. https://technostacks.com/blog/machine-learning-in-agriculture, last accessed on 28.08.2021.

[20] Saikat Mondal, Kanishka P. Wijewardena, Saranraj Karuppuswami, Nitya Kriti, Deepak Kumar, and Premjeet Chahal. Blockchain inspired RFID-based information architecture for food supply chain. *IEEE Internet of Things Journal*, 6(3):5803–5813, 2019.

[21] Nurzaman Ahmed, Debashis De, and Iftekhar Hussain. Internet of things (IoT) for smart precision agriculture and farming in rural areas. *IEEE Internet of Things Journal*, 5(6):4890–4899, 2018.

[22] Miguel Pincheira Caro, Muhammad Salek Ali, Massimo Vecchio, and Raffaele Giaffreda. Blockchain-based traceability in agri-food supply chain management: A practical implementation. In *2018 IoT Vertical and Topical Summit on Agriculture-Tuscany (IoT Tuscany)*, pages 1–4. IEEE, 2018.

[23] Jun Lin, Zhiqi Shen, Anting Zhang, and Yueting Chai. Blockchain and IoT based food traceability for smart agriculture. In *Proceedings of the 3rd International Conference on Crowd Science and Engineering*, pages 1–6, 2018.

[24] Tahmid Hasan Pranto, Abdulla All Noman, Atik Mahmud, and AKM Bahalul Haque. Blockchain and smart contract for IoT enabled smart agriculture. *PeerJ Computer Science*, 7:e407, 2021.

[25] Hang Xiong, Tobias Dalhaus, Puqing Wang, and Jiajin Huang. Blockchain technology for agriculture: Applications and rationale. *Frontiers in Blockchain*, 3:7, 2020.

[26] Sabir Hussain Awan, Sheeraz Ahmed, Asif Nawaz, S Sulaiman, Khalid Zaman, MY Ali, Zeeshan Najam, and Sohail Imran. Blockchain with IoT, an emergent routing scheme for smart agriculture. *International Journal of Advanced Computer Science and Applications*, 11(4):420–429, 2020.

[27] M. Shyamala Devi, R. Suguna, Aparna Shashikant Joshi, and Rupali Amit Bagate. Design of IoT blockchain based smart agriculture for enlightening safety and security. In *International Conference on Emerging Technologies in Computer Engineering*, pages 7–19. Springer, 2019.

[28] Malaya Dutta Borah, Vadithya Bharath Naik, Ripon Patgiri, Aditya Bhargav, Barneel Phukan, and Shiva GM Basani. Supply chain management in agriculture using blockchain and IoT. In *Advanced applications of blockchain technology*, pages 227–242. Springer, 2020.

[29] Feng Tian. An agri-food supply chain traceability system for china based on RFID & blockchain technology. In *2016 13th International Conference on Service Systems and Service Management (ICSSSM)*, pages 1–6. IEEE, 2016.

[30] Akash Suresh Patil, Bayu Adhi Tama, Youngho Park, and Kyung-Hyune Rhee. A framework for blockchain based secure smart green house farming. In *Advances in Computer Science and Ubiquitous Computing*, pages 1162–1167. Springer, 2017.

[31] Wen-Liang Chen, Yi-Bing Lin, Yun-Wei Lin, Robert Chen, Jyun-Kai Liao, Fung-Ling Ng, Yuan-Yao Chan, You-Cheng Liu, Chin-Cheng Wang, Cheng-Hsun Chiu, et al. Agritalk: IoT for precision soil farming of turmeric cultivation. *IEEE Internet of Things Journal*, 6(3):5209–5223, 2019.

[32] Agri Chain. https://agrichain.com/, last accessed on 02.09.2021.

[33] Agri Digital.https://www.agridigital.io/, last accessed on 02.09.2021.

[34] Agri Ledger.https://www.agridigital.io/, last accessed on 01.09.2021.

[35] IBM Blockchain. https://www.ibm.com/in-en/blockchain/solutions, last accessed on 05.09.2021.

[36] Agri10x. https://www.home.agri10x.com/, last accessed on 12.09.2021.

[37] Prince Waqas Khan, Yung-Cheol Byun, and Namje Park. IoT-blockchain enabled optimized provenance system for food industry 4.0 using advanced deep learning. *Sensors*, 20(10):2990, 2020.

[38] Sebastian Sadowski and Petros Spachos. Wireless technologies for smart agricultural monitoring using internet of things devices with energy harvesting capabilities. *Computers and Electronics in Agriculture*, 172:105338, 2020.

[39] Dilli Paudel, Hendrik Boogaard, Allard de Wit, Sander Janssen, Sjoukje Osinga, Christos Pylianidis, and Ioannis N. Athanasiadis. Machine learning for large-scale crop yield forecasting. *Agricultural Systems*, 187:103016, 2021.

[40] Jharna Majumdar, Sneha Naraseeyappa, and Shilpa Ankalaki. Analysis of agriculture data using data mining techniques: Application of big data. *Journal of Big Data*, 4(1):1–15, 2017.

[41] Rayda Ben Ayed and Mohsen Hanana. Artificial intelligence to improve the food and agriculture sector. *Journal of Food Quality*, 2021, 2021.

Index